CELL OF THE BLOOD

CELL OF THE BLOOD

ShivSanjeevi Sripathi

Delve
PUBLISHING

www.delvepublishing.com

Cell of the Blood

ShivSanjeevi Sripathi

Delve Publishing
4164 Lakeshore Road
Burlington, ON L7L 1A4
Canada
www.delvepublishing.com
Email: orders@arclereducation.com

ISBN: 978-1-77956-350-7 (Hardcover)

ABOUT THE AUTHOR

ShivSanjeevi Sripathi completed his Master's in Biotechnology from Mumbai University in 2008. He was awarded for academic excellence in both his Bachelors degree and Masters of Science for securing the second rank at the university level and first rank in his college. This was followed by a short foray into research on a stem cell project and later patent writing and immunoassays. This was followed by his entry into the teaching field where he taught biology and chemistry to 11th and 12th-grade students along with English grammar and communication skills to people of all age groups. He is an author/co-author of books on biotechnology/life sciences (more than 50 published books), which are in several international libraries. These books cover several topics like "depression: insights", "quantum biology", "viral epidemics of the 21st century", "yoga: research insights", "remote sensing in biology", "bionics", artificial intelligence in biology", molecular wires", "mitochondria", "stem cells", "nanotechnology", "camouflage" and "gene editing", to name a few. In his free time, he researches and gives talks on ancient Indian practices, temples and systems. This research compiles the science of traditional practices and how to interpret them in today's context. He has taught and interpreted several mantras and shlokas free of cost to a global audience as a service. He has appeared in a vernacular TV show (in Hindi) on a series of talks covering different temples and customs titled "Bharat ek punah khoj" (translates as searching again for the secrets of India) and on "Simha talks". His core belief is the ancient and powerful prayer "Tamaso maa jyotir gamaya" or "Lead us from darkness to the light".

TABLE OF CONTENTS

LIST OF FIGURES

1) as well as vascular cell adhesion molecule-1 (VCAM-1). Interleukin-6 (IL-6) and vascular endothelial growth factor (VEGF) were released in response to activation

Figure 2.5. RAGE isoforms. It is possible to produce AGE receptors as either complete molecules (Full-length, FL-RAGE) or as dominant negatives (Nt-RAGE) which is a non-efficient receptor. Soluble forms of RAGE can result from an incomplete transduction of the gene (esRAGE) or from a proteolytic reaction by ADAM 10 metalloproteinase

Figure 2.6. Binding of AGE to the endothelial cell receptor, RAGE. After this, the activation of NADPH oxidase and the formation of reactive oxygen intermediates are caused by a series of reactions involving diaphanous-1 (DIAPH 1), ERK1, ERK2, and protein kinase C. Through a pathway including the MAP kinase, NF-κB becomes activated and induces gene transcription

Figure 2.7. Blockade of RAGE activation. RAGE molecule has an extracellular part with two constant domains (pink circles) and one variable domain (blue octagon). By using anti-RAGE antibodies or synthetic peptide TTP488, AGE, amyloid protein, and S100 calgranulins can be inhibited from binding to the RAGE region (brown). In contrast, recombinant RAGE binds AGE ligands that compete against cellular RAGE. Endothelial cells were unable to transduce signals due to the blockade

Figure 2.8. (1) Leukocyte interactions and microdomain formation. As leukocytes roll and tether to PSGL-1, CD44 and ESL-1, they are bound by P- and E-selectins.LFA-1 and Mac-1 integrins are partially activated by these interactions, as well as Src kinase activity. The slow rolling is mediated by CD44 and ESL-1, which leads to the activation of neutrophils, which causes molecules to be polarized to the leading edge of the uropod. Different components of the blood interact with each other because intravascular leukocytes have activated adhesion receptors. PSGL-1 at the uropod and Mac-1 at the leading edge are putatively involved in interactions with platelets on polarized cells. Interactions between the leading edge and circulating RBCs are also mediated by the leading edge. This capture is mediated by ligands on platelets and RBCs that have not yet been identified. By promoting ROS release, engaged integrins deliver outside-in signals to adherent neutrophils. (2) Platelet interactions. In response to stimulated or damaged endothelium, resting or activated platelets can firmly arrest. Resting platelets express GPIbα and PSGL-1 which mediate the tethering on vWf and P-selectin on the endothelium, respectively. GPIIb/IIIa promotes aggregation of activated platelets or adhesion to vessel walls during inflammation. It has been shown that adherent platelets can enhance pathological conditions through P-selectin-mediated recruitment of leukocytes. Binding to leukocytes in the circulation is mediated by P-selectin and GPIbα, uncharacterized receptor/ligand pairs

Figure 3.1. An overview of white blood cells

LIST OF TABLES

PREFACE

"Eliminate the concept of division by class, skills, race, income, and nationality. We are all equals with a common pulse to survive. Every human requires food and water. Every human has a dream and desire to be happy. Every human responds to love, suffering, and pain. Every human bleeds the same color and occupies the same world. Let us recognize that we are all part of each other. We are all human. We are all one."

—Suzy Kassem

We have heard countless sayings in most languages about blood, such as "blood is thicker than water" or "my blood boils." In fact, the flow of blood in the radial artery or the pulse is what is checked first to determine if someone is alive! This pulse also provides insights into various aspects of our bodies when interpreted by an experienced healthcare practitioner. Additionally, the rush of blood to the face during moments of embarrassment, anger, or love is often described as "turning red," attributed to the hemoglobin in our blood.

In the ancient Sanatan Dharma Samskrutam texts, there are many comparisons to blood. For example, Ganesha is described as "*raktam*" (having a complexion that is red like blood), "*raktapushpaprpujitam*" (worshipped with flowers that are blood red, like a hibiscus) (*Ganapathy Atharvashirsha*). When considering the names of the divine mother Shakti, we find "raktavarna" (having a complexion that is red like blood), "*rudhirasamsthitha*" (seated in the blood as vitality) (*Lalita Sahasranama* or the 1000 names of *Lalitadevi, Brahm ndaPurana*).

The findings of several scientists like Hippocrates, Sushruta, Charaka, William Harvey, and Garvey, to name a few, set the path to understanding the essentials of the circulatory system. Current research has uncovered that the blood count or "blood profile" is indicative of one's health and can either aid in diagnosing a condition or show that "something is wrong" for further tests and diagnosis. Furthermore, a blood test is often a procedure that has become commonplace nowadays, especially with the recent COVID-19 pandemic.

"Everybody is a book of blood; wherever we're opened, we're red."

—Clive Barker, Books of Blood: Volumes One to Three

Many of us have learned the fundamental aspects of blood cells like the red blood cells, especially the role of hemoglobin. The white blood cells play an integral part in the defense against disease-causing microbes and cancer cells. The platelets help in clotting and without them, even a cut could be deadly. The altered profile of these blood cells in terms of their morphologies and counts illustrates what could be wrong in the body and can guide the appropriate treatment. Maybe what was said in the earlier times about "royal blood" is true, that one can get a picture of how someone is with their blood profile. This becomes even more pertinent given the emergence of new-age technology like sequencing and OMICS! Current diagnostics and future approaches aim at sequencing us right to our DNA using the cells of our blood!

"and so, I conclude that blood lives and is nourished of itself and in no way depends on any other part of the body as being prior to it or more excellent… So that from this we may perceive the causes not only of life in general… but also of longer or shorter life, of sleeping and waking, of skill, of strength and so forth."

—William Harvey, Disputations Touching the Generation of Animals (1651), trans. GwenethWhitteridge (1981), Chapter 52, 247.

This book covers the essentials of a few types of blood cells and goes on to discuss the number of these cells both under healthy conditions and during disease (can we say in sickness and health?). Results from research papers have been shared to enable a reader to see that the data is backed by actual findings.

An important aspect is then covered – blood transfusion. This becomes very important in today's scenario given the pandemic. In fact, the Red Cross issued a statement of dire shortage during the pandemic, which seems to have spilled over after the pandemic. Other aspects of which factors motivate one to donate blood and which factors deter from blood donation have also been covered. This can help plan appropriate systems to ensure that such a dreadful situation of blood shortage doesn't have to see the light of day again. An interesting section on "manufacturing blood" has also been covered. With continued research, maybe one day we can see such companies too. In line with this, a chapter on patient blood management (called PBM) has been covered discussing the ways to address such shortages.

Many of these chapters have included the relevance of artificial intelligence (AI) given the booming interest and emergence of this field. This becomes pertinent as AI is seeing applications in most sectors like mathematics, healthcare, science, materials science, medicine, geoscience, life science, physics, and chemistry, to name a few.

"Somewhere, something incredible is waiting to be known."

—Carl Sagan

Let us delve into select aspects of cells of the blood. There are so many new aspects of even already-known concepts awaiting along with an exciting future for research.

"We humans can never claim to do nothing, we breath, we pulse, we regenerate."
—Suzanne Weyn, Mr. Magorium's Wonder Emporium

Chapter 1

AN INTRODUCTION TO BLOOD CELLS-THEIR COUNT IN SICKNESS AND HEALTH

"The idea of Dr. Linus Pauling that an abnormal hemoglobin molecule might be responsible for the sickling process initiated the study of the hemoglobin molecule in hereditary anemias."
—Harvey Itano. "Clinical States Associated with Alterations of the Hemoglobin Molecule." Archives of Internal Medicine, 96: 287–97, 295. 1955.

1.1. SMEARS

A complete blood count (CBC) test is the most fundamental blood test to measure the different basic components of blood. The varying blood cell count from the test aids in detecting and analyzing an underlying health condition. It aids in making a proper diagnosis of one's health. The main types of blood cell count that aid in determining the medical condition comprise red blood cells (RBCs), white blood cells (WBCs), platelets, and plasma, and each of these has specific functions. RBCs deliver oxygen to tissues, WBCs are associated with immunity by lymphocytes and ingestion and breaking down of microbes and foreign particles by granulocytes and monocytes, while platelets are involved with clot formation. Circulation of blood acts as a channel that accumulates different kinds of cells to the regions in the body where they are needed, for example: RBCs to supply oxygen to tissues, platelets to the sites of injury, phagocytic cells to sites of inflammation and those invaded by microbes, and lymphocytes to infectious tissues (Schwartz et al., 2023) (Figures 1.1 and 1.2).

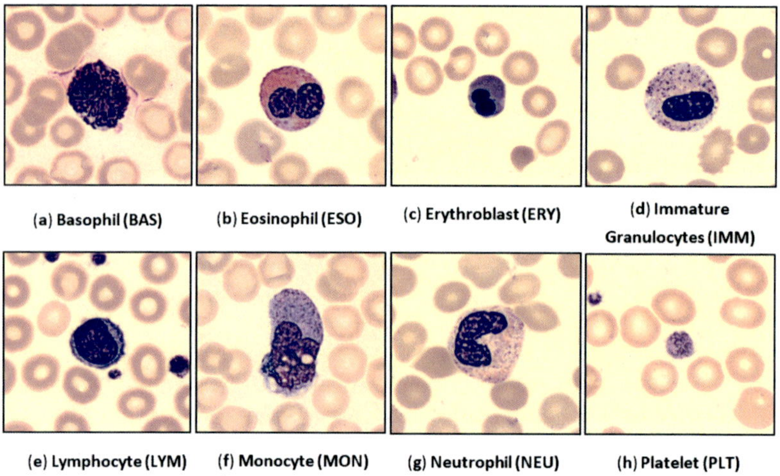

Figure 1.1. *Examples of blood smears.*

Source: Chola, C., Muaad, A. Y., Bin Heyat, M. B., Benifa, J. V. B., Naji, W. R., Hemachandran, K., Mahmoud, N. F., Samee, N. A., Al-Antari, M. A., Kadah, Y. M., et al., (2022). BCNet: A deep learning computer-aided diagnosis framework for human peripheral blood cell identification. Diagnostics, 12(11), 2815. *https://doi.org/10.3390/diagnostics12112815: © 2022 by the authors. Licensee MDPI, Basel, Switzerland. This article is an open access article distributed under the terms and conditions of the Creative Commons Attribution (CC BY) license: https://creativecommons.org/licenses/by/4.0/).*

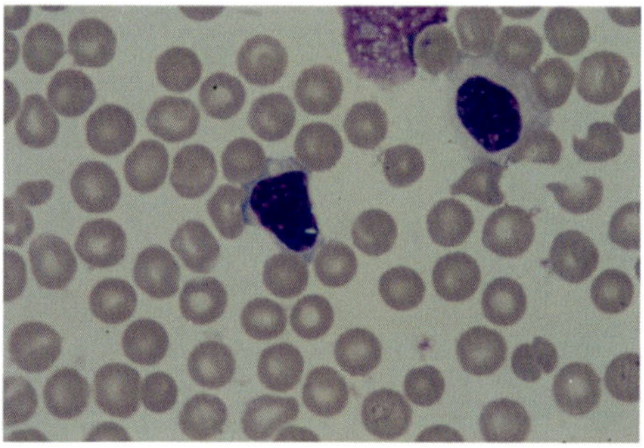

Figure 1.2. Examples of blood smears

Source: Dralus, G., Mazur, D., &Czmil. A., (2021). Automatic detection and counting of blood cells in smear images using retinanet. Entropy, *23(11):1522. https://*

1.2. RED BLOOD CELLS (RBCS)

These cells are also known as erythrocytes. They are the oxygen carriers carrying oxygen from the lungs to the whole body and transport back carbon dioxide from the body to the lungs for exhalation (Barbalato&Pillarisetty, 2022). A low RBC count may point out an underlying health condition like anemia (Turner et al., 2023) or a high RBC count may suggest polycythemia vera (Lu & Chang, 2023). Along with oxygen and carbon dioxide, they also carry blood group antigens and proteins like the Rh factors that indicate the blood type of an individual. RBCs also act as buffers for the regulation of hydrogen ion concentration. They contribute to the blood viscosity.

"The hemoglobin molecule, with its striking color and its property of combining reversibly with dioxygen, seemed to me to be especially interesting."

Linus Pauling. "Hemoglobins and Hemoglobinopathies: A Current Review to 1981." Texas Reports on Biology and Medicine, University of Texas Medical Branch, Galveston, Texas, 40: 1–7. 1980.

Mature RBCs are anucleated, small and biconcave disk-shaped, roughly dumbbell-shaped and approximately 7.8 μm in diameter. The biconcave structure confers them a larger surface-to-volume ratio aiding in sufficient gaseous exchange thus enabling the cells to carry out their functions efficiently. This structure enables the flexibility essential for the passage through the cardiovascular system. Owing to its flexible structure, when necessary, as in the case of movement through tiny blood vessels, it can reform taking a bell shape to pass through effortlessly. The cells have a phospholipid bilayer membrane that is maintained by a network of proteins helping in making up the cytoskeleton. The cytoskeleton is made of actin, protein 4.1, spectrin, ankyrin, actin and band 3 which permits cellular malleability and structural integrity (Barbalato&Pillarisetty, 2022).

1.2.1. Function of RBCs

The lifespan of an RBC is approximately 120 days. Throughout this time frame, it has to deliver oxygen from the lungs to all the tissues and cells to assist metabolic pathways like ATP synthesis. It also carries carbon dioxide generated from the cells and tissues back to the lungs for waste elimination through exhalation. Deoxygenated blood (rich in carbon dioxide) contains hemoglobin with Fe (ferrous heme) that has an affinity for oxygen. When blood reaches deoxygenated tissues, there is a low pH and reduced partial pressure of oxygen. This causes the heme to lose its oxygen affinity and deliver it to the tissues. Subsequently, carbon dioxide enters the RBC combined with water, forming hydrogen and bicarbonate by carbonic anhydrase. The majority of the carbon dioxide travels to the lungs in this bicarbonate form, which is then expelled from the system through exhalation (Barbalato&Pillarisetty, 2022) (Figure 1.3).

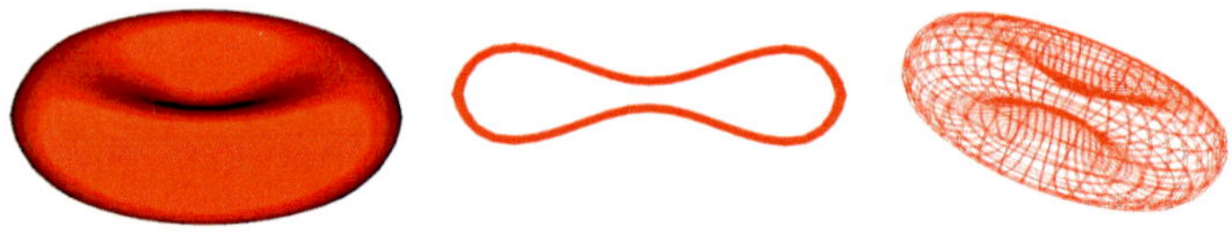

(a) 3D RBC model (b) Cross section (c) Surface mesh

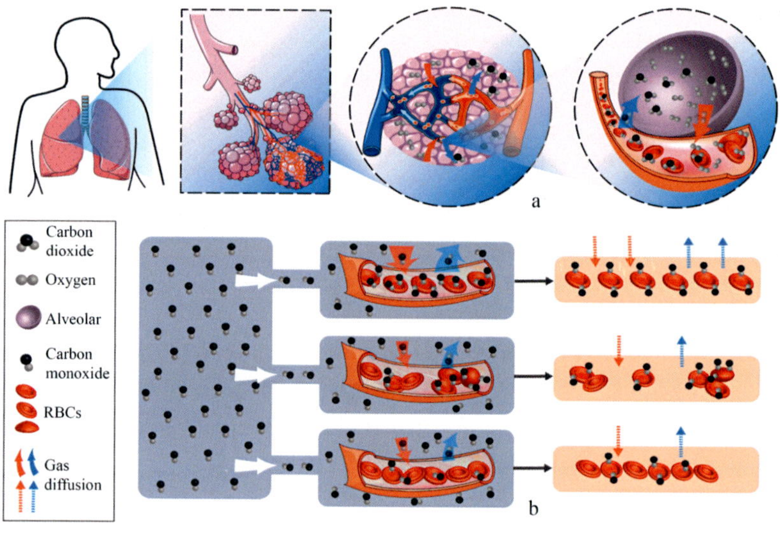

Figure 1.3.*A study that modeled RBCs in terms of their 3D shape and functioning.*

Source: An, L., Ji, F., Yin, Y., Liu, Y., & Zhou, C., (2022). Modeling of red blood cells in capillary flow using fluid–structure interaction and gas diffusion. Cells, 11(24):3987. https://doi.org/10.3390/cells11243987: © 2022 by the authors. Licensee MDPI, Basel, Switzerland. This article is an open access article distributed under the terms and conditions of the Creative Commons Attribution (CC BY) license (https://creativecommons.org/licenses/by/4.0/).

RBCs can be analyzed by electron and light microscopy. Each of the techniques reveals specific details of their structure. Under the microscope, RBCs appear as yellow-green disks and their centers are pale. The volume of packed RBCs (hematocrit value) is obtained by centrifugation of the sample, and the hematocrit value ranges between 37–47% in women, 42–54% in men, and lower values in children. Healthy and normal RBCs have a uniform value, and hence the hematocrit value is fixed by the number of RBCs per unit of blood. The normal RBC count varies from 4 million to 6 million per cubic millimeter. Histochemical stains, eosin, and hematoxylin are also known to implement essential functions in determining the essential elements in the RBCs while helping in the diagnosis of diseases. Wright stains are used for observing Howell-Jolly bodies, which are the fundamental component in erythroblastosis, megaloblastic anemia, hyposplenism, myelodysplasia, and post-chemotherapy. Giemsa stains, on the other hand, are used for the diagnosis of lymphoma and distinguishing plasmodium of malaria (Barcia, 2007).

1.3. WHITE BLOOD CELLS (WBCS)

WBCs, also known as leukocytes, comprise lymphocytes and phagocytic cells (monocytes and granulocytes). They hold the defense mechanism of the body, helping the body to ward off infections and disease-causing microorganisms, and are also involved in the repairing activity. They take part in both humoral and innate immune responses (Tigner et al., 2022). Depending on the presence or absence of granules in the cytoplasm, the WBCs are differentiated as granulocytes and agranulocytes. They are nucleated and motile and do not undergo mitosis; however, some of them preserve the capability of undergoing mitosis. They are highly and specifically differentiated as per their specialized functions. The normal number of WBCs in blood samples varies from 4,500 to 11,000 per cubic millimeter. Variation is observed in this range during the day, where the values are low in resting conditions and higher during exercises. If the physical exertion is intense, the normal count can exceed 20,000 per cubic millimeter. The majority of the WBCs are not a part of the circulation, and very few are present in the blood and are journeying within different sites. Being nucleated and due to their ability to produce ribonucleic acid (RNA), WBCs can synthesize proteins. WBCs are further classified into three categories, namely monocytes, granulocytes, and lymphocytes, where each of them is varied in their structure and function (Schwartz et al., 2023).

- **Granulocytes:**Mainly of three types: eosinophils, basophils, and neutrophils. They consist of azurophilic granules called lysosomes along with specific granules that hold particles specific to each of the cell's functions. Based on the morphology of their size, nucleus, and pattern in which the granules stain, granulocytes can be easily distinguished (Prinyakupt&Pluempitiwiriyawej, 2015).

- **Neutrophils**: Polymorphonuclear, meaning they have multiple-lobed nuclei comprising three to five segments amalgamated by isthmuses or thin strands, and they vary in size from 12–15 µm in diameter. Their lifespan is of a few days and on activation, they experience apoptosis and later are removed by macrophages in the connective tissues. They resemble scavengers that encircle fungi and bacteria present in the body.
- **Eosinophils:** Consist of a bi-lobed nucleus and are the same size as neutrophils but possess larger granules, and the chromatin is usually saturated in only two non-segmented lobes of the nucleus. They migrate from the blood circulation to sites in the skin, respiratory tract, and lungs via the lymphatic system. Eosinophils are phagocytic and actively motile, offering defense against parasites. They contribute to inflammatory, immune, and hypersensitivity reactions.
- **Basophils:**comprise either S-shaped nuclei or are bi-lobed. Their size varies between 12–15 µm in diameter and they contain granules of sulfated glycosaminoglycans and heparin. Post-release from bone marrow, they navigate from circulation to the skin and mucosa which form the barrier tissues. Along with histamine, they cause a hypersensitive reaction along with neutrophils, macrophages, and platelets. Their number increases upon having an allergic reaction (Tigner et al., 2022) (Figure 1.4).

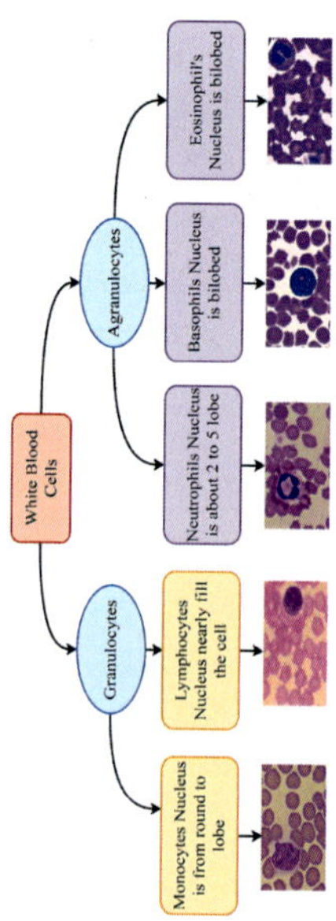

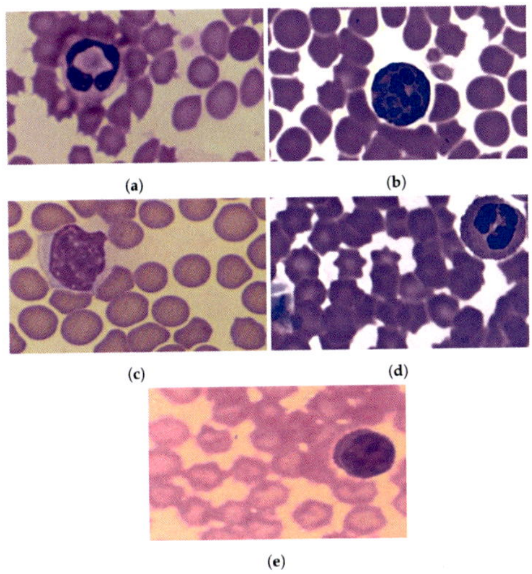

Figure 1.4. *The types of WBCs.*

Key: (a) Neutrophils, (b) Basophils, (c) Eosinophils, (d) Monocytes, and (e) Lymphocytes

Source: Rustam, F., Aslam, N., De La Torre Díez, I., Khan, Y. D., Mazón, J. L. V., Rodríguez, C. L., & Ashraf, I., (2022). White blood cell classification using texture and RGB features of oversampled microscopic images. Healthcare, 10(11):2230. https://doi.org/10.3390/healthcare10112230: © 2022 by the authors. Licensee MDPI, Basel, Switzerland. This article is an open access article distributed under the terms and conditions of the Creative Commons Attribution (CC BY) license (https://creativecommons.org/licenses/by/4.0/).

1.3.1. Agranulocytes

Monocytes and lymphocytes are the two types of agranulocytes present in humans where the cell cytoplasm is devoid of granules. Around 35% of the hematologic blood values are agranulocytes.

1.3.1.1. Monocytes

Structurally, they are the largest blood cells with their average size ranging between 15–18 μm in diameter, and they also constitute around 4–8% of the total WBCs. The nucleus is comparatively larger in size and is folded or indented. The cytoplasm holds a large number of fine particles which are dense in close proximity to the cell membrane. They are active, motile, and phagocytic in nature. They form the precursor cells for microglial cells, osteoclasts, and macrophages in organs and especially connective tissues. Monocytes are proficient in ingesting RBCs, large particles, and infectious particles; however, they cannot be substituted for the removal and destruction of bacteria, which is done by the neutrophils. They follow granulocytes in inflamed areas and are usually located at the sites of chronic infections. Navigating through bone marrow, they enter tissues developing into macrophages that are present in almost all tissues of the body. In the skin, they are called Langerhans cells, while in the liver, they are named Kupffer cells. Being excellent scavenger cells, they increase immunity response by digesting and processing antigens, making them easily distinguished by lymphocytes as foreign particles (Häggström, 2014). An illustration of WBC reference ranges differentiating mononuclear WBC (in yellow) vs. other cells is presented Figure 1.5.

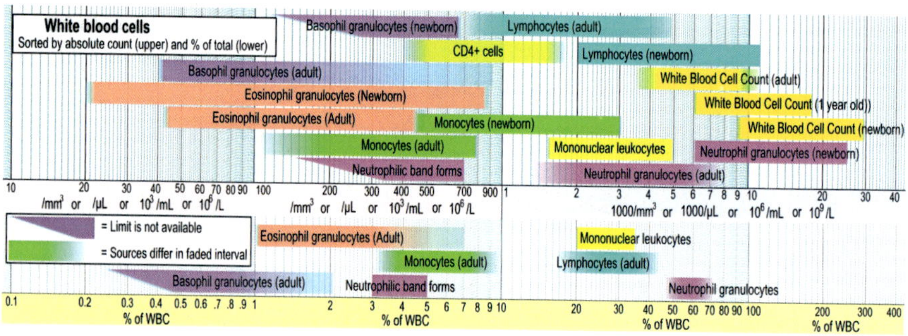

Figure 1.5. *An illustration of WBC reference ranges differentiating mononuclear WBC (in yellow) vs. other cells.*

Source: Häggström, (2014). "Medical gallery of Mikael Häggström 2014." Wiki Journal of Medicine 1(2): "All these medical images are public domain under the creative commons CC0 1.0 universal public domain dedication which means that you can copy, modify, distribute and perform the works, even for commercial purposes, all without asking permission."

1.3.1.2. Lymphocytes

They form part of the immune response that the body mounts against foreign particles. Their size varies; they have spherical nuclei, and around 23–43% of leukocytes are lymphocytes. The larger activated lymphocytes range between 9 to 18 µm in diameter, have indented nuclei, and cytoplasm rich in azurophilic granules. The smaller ones are approximately the size of RBCs, with scantier cytoplasm and spherical heterochromatic nuclei. Using CD markers for differentiation, they are mainly grouped into B and T lymphocytes. They are predominantly present in lymph nodes, thymus, lymphoid tissue of the GIT (gastrointestinal tract), tonsils, and spleen. They travel through lymphatic channels and flow into the thoracic lymph duct, which is connected to the venous circulatory system. Their lifespan is about a year or more, and unlike other blood cells, they may leave and re-enter circulation, with their main route being through lymph nodes or the spleen. They can be activated and stimulated to undergo mitosis when triggered by antigens or other agents. They are responsible for reactions when invaded by foreign microbes and cells, such as those in transplanted organs and foreign proteins.

Their two classes are recognized by the immune response induced by them. B cells/lymphocytes participate in humoral immunity where they secrete immunoglobulin or antibodies when encountered by foreign particles. T cells/ lymphocytes, on the other hand, participate in cell-mediated responses where they directly attack foreign particles and manage the antibody formation function exhibited by the B cells. They are also responsible for some allergic reactions and rejecting transplanted tissues. The thymus, a small organ behind the breastbone, is comparatively large at birth but shrinks post-puberty and in the elderly, it remains as a fibrous cord. Before birth, the thymus initiates its role in differentiating lymphocytes and hence, if removed at birth, it hinders the development of immunological responses. Mature T lymphocytes migrate from the thymus through circulation to the spleen and lymph nodes. B and T cells function in coordination with each other where the T cells are always in surveillance mode for detecting cell surfaces for foreign antigens and influence the B cells for their next move on destruction of the foreign antigens, for example, immunity against viruses (Tigner et al., 2022).

1.3.2. Platelets

Platelets are also known as thrombocytes and are the tiniest cells that help in the clotting of blood at the site of an injury. Their size varies between 2–4μm in diameter and the number ranges between 150,000 to 400,000 per cubic millimeter. Although compared to WBCs, they are present in copious amounts, they hold a comparatively smaller fraction of blood volume due to their minute size. They too, like the RBCs, are anucleated and hence incapable of mitosis (cell division). However, they exhibit a much more complex internal structure and metabolic pathway compared to the RBCs. In fresh blood samples, they are seen as spheroid yet they exhibit the tendency to exude hairlike filaments from their membranes. They are observed to be adhering to each other and not to RBCs or WBCs. The tiny granules in platelets possess materials that are essential for clot formation, the main function of the platelets.

Their function is associated with hemostasis, which is the control and prevention of bleeding. In case of injury to the endothelial lining of any blood vessel, the platelets instantly attach to the site of injury and among themselves in large numbers constructing an adherent mass of platelets. Their function and response is to stop bleeding and form a thrombus or a blood clot. In their absence, this essential defense mechanism is curbed, which results in prolonged bleeding from wounds big or small. Platelets are responsible for the prevention of leakage of RBCs by resisting the capillary membranes. Deficiency of platelets results in the scaling down of resistance of capillary walls, which further causes unusual bleeding from the capillaries that happens as a result of a minor injury or can occur instantaneously. They are responsible for providing materials for normal blood coagulation while also resulting in retraction or shrinking that ultimately results in clot formation.

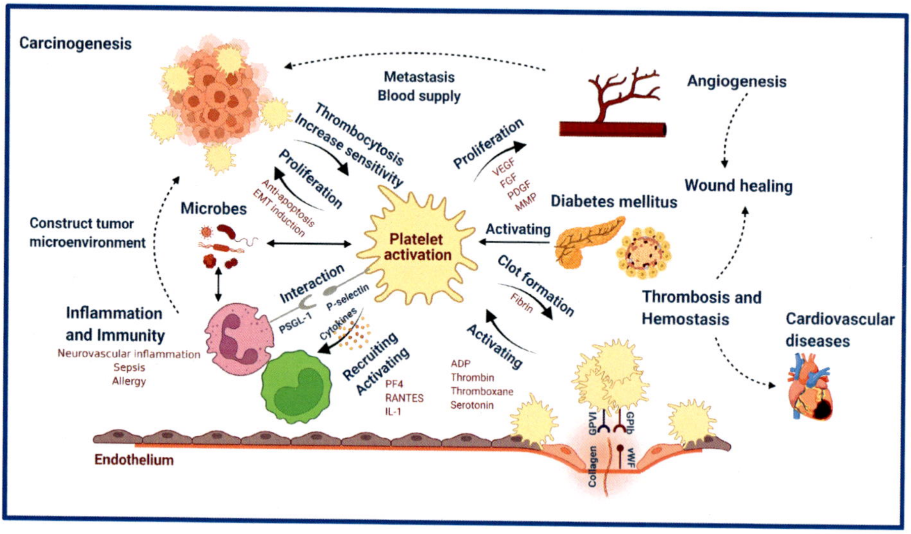

Figure 1.6. *Illustrates the versatility of platelet function in health and disease.*

Source: Chaudhary, P. K., Kim, S., & Kim, S., (2022). An insight into recent advances on platelet function in health and disease. International Journal of Molecular Sciences, 23(11), 6022. https://doi.org/10.3390/ijms23116022: © 2022 by the authors. Licensee MDPI, Basel, Switzerland. This article is an open access article distributed under the terms and conditions of the Creative Commons Attribution (CC BY) license (https://creativecommons.org/licenses/by/4.0/).

Their lifespan is approximately ten days when released into circulation and then later removed and destroyed by tissue phagocytes, also known as reticuloendothelial cells.

Their production rate is not as controlled as compared to that of RBCs, and the hormone thrombopoietin is considered to be the chemical mediator for regulating their numbers in the bloodstream. Except for the spleen, there are no reserves of platelets in the body, and they are numerous in the spleen compared to their concentration in the peripheral blood. An abnormal platelet count indicates conditions that lead to bleeding disorders or an elevated risk of uncontrolled excessive clotting (Tigner et al., 2022) (Figure 1.6).

1.4. CLINICAL SIGNIFICANCE

A complete blood count (CBC) test is a laboratory examination that measures the various blood particles present in a given blood sample. It is the most basic and first line of diagnostic tests to guide in making a proper diagnosis of a patient's health.

From the detailed report, we learn the range of the RBCs, WBCs (neutrophils, eosinophils, etc.), platelets, etc., as a percentage. Comparing it with the normal range of these values, one can make clinical inferences of their values to be elevated or depleted and aid in getting a direction of the possible health condition of a patient; for example, a lower RBC value may indicate anemia.

The reference range of the count of blood cells varies marginally based on the laboratory and the testing techniques used by them. Below is the general range of the blood cell count which is considered to be normal or within the healthy range; anything above or below is considered a matter of an underlying health condition which leads to further tests to find the exact disease. It helps understand the cause behind symptoms of weakness, fever, and tiredness. It also helps in understanding the reason behind bleeding, swelling, bruising, and pain. A CBC test aids in the surveillance of the factors that cause a particular health condition.

Post-diagnosis, a CBC helps in determining the impact of the given treatment on health conditions and makes way for continuing with the same treatment or choosing a different mode of treatment. For example, in the case of cancer treatments, looking at the blood cell count, the physician can determine the impact of the given treatment and how it affects the patient's health with medicine and radiation.

From the table, anything above or below the range of CBC is considered an underlying health condition.

For example, any value above the mentioned WBC count is considered as leukocytosis and any value below the range is considered as leukopenia. However, further tests to confirm the findings are needed as these values point towards the probability of the underlying health concern.

Table 1.1. The Normal Range of Expected Complete Blood Count Results for Adults Which is Sex and Age-Specific in Nature

Test Acronym	Normal Range Values (Male)	Normal Range Values (Female)
WBC	4.0–10.8 × 103/μL	4.0–10.8 × 103/μL
RBC	4.5–6.1 × 106/μL	4.0–5.4 × 106/μL
Hb	13.0–17.0 g/dL	12.0–16.0 g/dL
HCT	40.0–52.0%	37.0–47.0%
MCV	80–98 fL	80–98 fL
MCH	27.0–33.0 pg	27.0–33.0 pg
MCHC	31.5–37.0 g/dL	31.5–37.0 g/dL
RDW	11.5–14.5%	11.5–14.5%
PLT	150–400 × 103/μL	150–400 × 103/μL
Neutrophils %	40–73%	
Lymphocytes %	19–48%	
Monocytes %	0.4–10.0%	
Eosinophils %	0–7.0%	
Basophils %	0–2.0%	

Key: hemoglobin (Hb), mean corpuscular volume (MCV), mean corpuscular hemoglobin (MCH), mean corpuscular hemoglobin concentration (MCHC), red cell distribution width (RDW), mean platelet volume (MPV), neutrophil-to-lymphocyte ratio (NLR), platelet-to-lymphocyte ratio (PLR), and monocyte-to-lymphocyte ratio (MLR)

Source: Seo I. H., & Lee Y. J., (2022). Usefulness of complete blood count (CBC) to assess cardiovascular and metabolic diseases in clinical settings: A comprehensive literature review. Biomedicines, 10(11), 2697. https://doi.org/10.3390/biomedicines10112697:

- **CBC in Sickness:** From the above table, we learned that the values of CBC tests, if above or below, point towards certain health conditions. Hereafter, we will look into their values in case of certain health conditions that could be caused by bacteria or viruses or be an autoimmune disorder.

- **Changes in RBC Count:** Any changes observed in the reports in the values of RBCs contribute to the essential diagnostic information for several health conditions and diseases. Below are some examples of how changes in the values of RBCs help in the diagnosis of these conditions.

1.4.1. Anemia

It is a health condition that is indicated by a significant decrease in the number of RBCs; hematocrit (HCT) or hemoglobin (Hb). Anemic patients exhibit symptoms of weakness, lethargy, and extreme fatigue. Severe cases of anemia may exhibit symptoms of shortness of breath, chest pain, restless legs, syncope, and decreased tolerance to exercise. Also, the skin may be cool to the touch. It is not the diagnosis but an important presentation of an underlying health condition. The symptomatic nature of the condition highly depends on the etiology of anemia, the occurrence of other co-existing conditions like the presence of cardiovascular disease, and its sensitivity to onset. The majority of the patients encounter its symptoms when the Hb levels are lower than 7.0g/dL.

Nutritional supplements of Vitamin B12, iron, and folate are to be prescribed. In the case of iron deficiency, the supplements are advised to be consumed for three months after the levels are shown to be within the normal range. If left undiagnosed and untreated for a prolonged time frame, it can result in multiple organ failure and death. Expecting mothers can undergo premature labor, and the babies may be born underweight. Anemia in pregnancy can result in a large amount of blood loss during pregnancy and exponentially escalate the chances of the baby being born with anemia (Santoscoy et al., 2013).

Table 1.2. Apart from Variation Dependent on Laboratory Techniques, the Normal Hb (Turner et al., 2023)

In men 13.5 to 18.0 g/dL In women 12.0 to 15.0 g/dL In children 11.0 to 16.0 g/dL The values are varied in pregnancy based on the trimester, however, majorly they are greater than 10.0 g/dL.

The older population faces more complications in chronic anemia where mainly their cardiovascular health is severely impacted. High-output heart failure, myocardial infarction, cardiac hypertrophy, arrhythmias, and angina are the commonly known complications (Triscott et al., 2015).

Severe cases of anemia from a younger age may result in compromised neurological development impacting areas of mental, cognitive, and delays in overall growth. These cannot be easily managed medically (Turner et al., 2023) (Table 1.2).

There is a variety of anemia caused by causes like:

- **Anemia Due to Iron-Deficiency (MCV < 80 fl):** Here, the RBCs are microcytic (smaller in size) and contain less hemoglobin due to a shortage of iron for the synthesis of hemoglobin. The mean corpuscular value (MCV) is < 80 fl and can lead to a dimorphic cell population.
- **Anemia due to Vitamin B12 Deficiency (MCV > 100 fl):** Here, the RBCs are macrocytic (larger in size) and can be oval in shape as a result of impaired DNA synthesis that happens in the bone marrow.
- **Hemolytic Anemia (MCV 80–100 fl):**Here, the RBCs are destroyed prematurely which causes a low RBC count, an increase in young RBCs also known as reticulocyte count and buildup of bilirubin (Turner et al., 2023).

Table 1.3. Summarizes the Differences Between iron Deficiency Anemia and Anemia in Chronic Disease

Feature	**Iron Deficiency Anemia**	**Anemia of Chronic Disease**
The presence of other chronic diseases	Rarely	Often
Average baseline hemoglobin concentration	≥9 g/dL	≤9 g/dL
Serum iron concentration	Decreased	Significantly decreased
MCV/MCH	Decreased	Normal
Serum ferritin concentration	Low	Increased
Serum hepcidin concentration	Low	High
Percentage of reticulocytes in the blood serum	High	How

Serum folic acid concentration	Normal	Decreased
Serum vitamin B_{12} concentration	Normal	Decreased
Serum creatinine concentration	Normal	Increased
Serum erythropoietin concentration	Increased	Decreased

Source: Wiciński M, Liczner G, Cadelski K, Kołnierzak T, Nowaczewska M, Malinowski B., (2020). Anemia of chronic diseases: wider diagnostics—Better treatment? Nutrients, 12(6), 1784. https://doi.org/10.3390/nu12061784: © 2020 by the authors. Licensee MDPI, Basel, Switzerland. This article is an open access article distributed under the terms and conditions of the Creative Commons Attribution (CC BY) license (http://creativecommons.org/licenses/by/4.0/).

1.4.2. Polycythemia

This health condition is identified by unusually increased levels of RBCs. This could be either primary as in the case of polycythemia vera or secondary as in the case of situations where the response is adapted due to low oxygen levels as observed in lung diseases. The normal range of RBC mass depends on the altitude, country, and ethnicity of the individual, and the reference frame does not exceed 32 ml/kg in women and 36 ml/kg in men. Polycythemia in newly born babies is elucidated as a Hb value above 22g/dL or central venous hematocrit over 65% (Pillai et al., 2023).

Common symptoms are headache, fatigue, blurry vision, transient ischemic attacks, a history of gastrointestinal bleeding, epistaxis, and easy bruising.

Progression to leukemia, particularly acute myeloid leukemia (AML), is seen in approximately 5% of cases and is often refractory to treatment. Studies have implicated the use of chlorambucil, pipobroman, or radioactive phosphorus as factors that increase the likelihood of progression (Pillai et al., 2023).

1.4.3. Autoimmune Hemolytic Anemia (AIHA)

In this condition, the person's immune system attacks and destroys the red blood cells resulting in anemia. This can be diagnosed by signs of hemolysis and a high reticulocyte count. The Hb levels are more than 7.5 g/dl or 70.5 g/L (Hill & Hill, 2018). A summary of findings in patients with primary AIHA is shown in Figure 1.7.

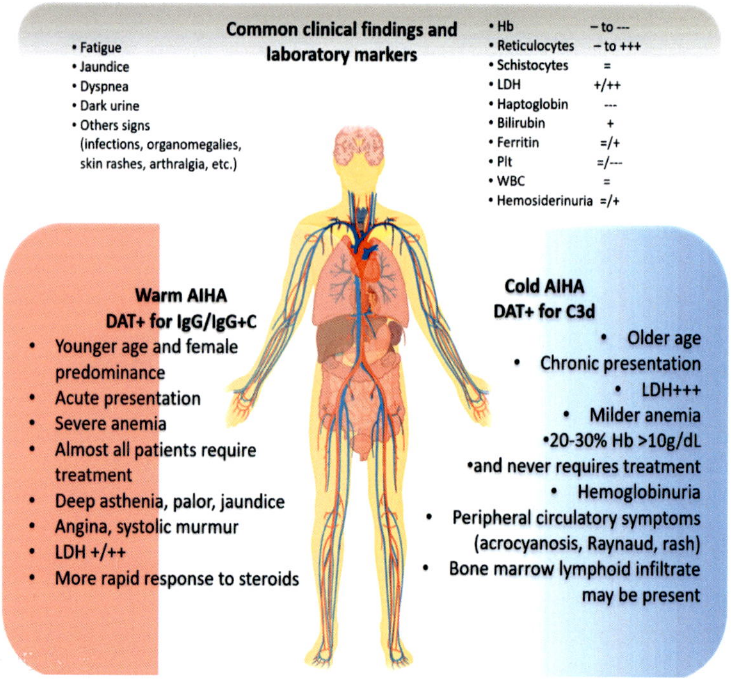

Figure 1.7. *A summary of findings in patients with primary AIHA.*

Key: warm (wAIHA) and cold (cAIHA), lactate dehydrogenase LDH, normal =, slightly reduced -, reduced --, and very reduced --- or slightly increased +, increased ++, and very increased +++, and direct antiglobulin test (DAT).

Source: Fattizzo B, Giannotta JA, Serpenti F, Barcellini W., (2020). Difficult cases of autoimmune hemolytic anemia: a challenge for the internal medicine specialist. Journal of Clinical Medicine 9(12), 3858. https://doi.org/10.3390/jcm9123858: © 2020 by the authors. Licensee MDPI, Basel, Switzerland. This article is an open access article distributed under the terms and conditions of the Creative Commons Attribution (CC BY) license (http://creativecommons.org/licenses/by/4.0/).

1.4.4. Aplastic Anemia

In this condition, the bone marrow doesn't make enough quantity of blood cells including RBCs. This results in a lower RBC count and most of the time, it is the cause of extreme fatigue and other symptoms. It is found to occur in young patients with pancytopenia. The symptoms linked with the absent cell lineage are severe anemia that progresses to weakness and fatigue along with dyspnea, thrombocytopenia, mucosal bleeding, ecchymoses, petechiae, and frequent minor infections. The incidences are 0.6–6.1 cases per million of the total population with equal distribution in genders and race, and this figure is established from the data procured from retrospective reviews of death registries (Moore and Krishnan, 2023).

Table 1.4. Compounds implicated in aplastic anemia

Drugs	
•	*Antibiotics*
o	Chloramphinicol, Sulphonamide, cotrimoxazole, linezolid.
•	*Antirheumatics*
o	Gold salts
o	Hydroxychloroquine
o	Allopurinol
o	Penicillamine.
•	*Anticonvulsants*
o	Phenytoin, carbamazepine.
•	*Anti-inflammatory*
o	Piroxicam, indomethacin, diclofenac, sulfasalazine, naproxen.
•	*Antithyroid*
o	Carbimazole, thiouracil.
Chemicals	
•	*Benzene*
•	*Pesticides*
o	Organochloride, organophosphates.
•	*Narcotic drugs*
o	Ecstasy, MDMA, methamphetamine.

Source: Solimando AG, Palumbo C, Pragnell MV, Bittrich M, Argentiero A, & Krebs M., (2022). Aplastic anemia as a roadmap for bone marrow failure: An overview and a clinical workflow. International Journal of Molecular Sciences, 23(19), 11765. https://doi.org/10.3390/ijms231911765:

Severe aplastic anemia is determined by the hypocellularity of bone marrow which is < 25%. The superlative condition is determined by the following parameters: absolute neutrophil count < 200/microL (< 0.2 x 109/L). Absolute neutrophil count < 500/microL (< 0.5 × 109/L)

Platelet count < 20,000/microL (< 20 × 109/L)

Absolute reticulocyte count < 60,000/microL (< 60 × 109/L) (Biswajit et al., 2012).

Interestingly, a research article compiled several compounds that have been implicated in causing this condition: they have been presented in the Table 1.14.

1.4.5. Sickle Cell Disease

It is a genetic disorder where the RBCs take a crescent or sickle shape while becoming rigid, and this happens when the oxygen levels are considerably low. Due to the sickle shape, it could lead to blocking of blood vessels, which causes pain and organ damage if the tissues don't receive proper.

Table 1.5. A review of techniques to detect sickle cell disease (*the references for each technique can be accessed from the main citation*)

Technique	Sensitivity	Specificity	Accuracy	Advantages	Disadvantage	Result
Peripheral blood smear (PBF)	35.0%.	96.7%	90.5%	Simple preparation, inexpensive, Turnaround time (TAT) is 44 min	Dependence on the pathologist's skills, does not differentiate between different types of SCD	Detect sickle cells
Solubility and Sickling	Sickling: 65.0% Solubility: 45.0%.	Sickling: 95.6% Solubility: 90.0%.	Sickling: 92.5% Solubility: 85.5%.	Easy, inexpensive, fast, affordable, TAT 38 min for sickling, TAT for solubility 70 min	Testing newborns shows false-negative result, does not differentiate between SCD types	Detect the sickling event.

Capillary electro-phoresis	Not reported	Not re-ported	Not reported	Reliable, ability to distinguish most types of sickle cell disease including heterozy-gous.	Expensive, requires skilled technicians	iden-tify and quantify HbF, Hb A, Hb A2, Hb S, Hb C, Hb Barts and other
Iso-electric focusing (IEF)	Not reported	Not re-ported	Not reported	Detect HbS and HbA easily in a high con-centration of HbF, Hb D-Punjab easily separated from HbS, need a small volume of the sample, able to use dried blood spot, TAT is 45 min.	Expensive, requires highly trained staff to in-terpret the results.	Hb A, Hb F, Hb C, Hb S, Hb E and Hb O Arab
High-perfor-mance liquid chroma-tography (HPLC)	Not reported	Not re-ported	Not reported	Reliable, ability to distinguish most types of sickle cell disease including heterozy-gous, fully automated	Misdi-agnoses the new variants that mimic HbS, Ex-pensive and needs trained person-nel, not practical in limited resources areas	Detect Hb F, Hb A2, Hb S, Hb C, Hb Barts, and other Hb vari-ants.
Amplifi-cation-re-fractory mutation system (ARMS) poly	75%	Not re-ported	Not reported	Simple, can be used for prenatal diagnosis	Low sensitivity, maternal cell DNA contamina-tion	$\beta S\beta S$ $\beta A\beta S$ $\beta A\beta A$

merase chain reaction (PCR) for prenatal analysis						
Allele-Specific Recombinase Polymerase Amplification	100%	βA: 94.7% βS:97.1%	<95%	Affordable, rapid (less than 30 min), low-cost, accurate	This test is difficult to design, missing some single nucleotide polymorphisms (SNPs), costly and laborious assay	βA βS
Image processing technique	96.55%	Not reported	95%	Automated method to detect sickle cells, minimize the error of dependence on the naked eye	Cannot distinguish between different types of SCD, cannot be used to determine the severity of the disease, affected by different conditions that can affect the red blood cell (RBCs) numbers as in blood transfusion, expensive, needs special equipment such as camera connected to microscope	Detect Sickling RBCs

Propose deep learning models	Not reported	Not reported	99.54%	Indicate the sickle RBCs automatically in one shot, minimize the error of dependence on the naked eye	Cannot distinguish between different types of SCD, cannot be used to determine the severity of the disease, affected by different conditions that can affect the RBCs number as in blood transfusion, needs special equipment such as a camera connected to the microscope., time consuming, ignore other cells which leads to false diagnosis	Detect Sickling RBCs
Smartphone microchip with microscope and machine learning algorithms	Not reported	Not reported	Not reported	Can be used as a point of care (POC) to monitor the disease severity, reduce the cost	Test is based on the morphology of the RBCs, cannot distinguish between different types of SCD, affected by different conditions that can affect the RBCs morphology	Detect Sickling RBCs

Electrical impedance microflow cytometry	91%	86%	Not reported	Used to monitor the sickling events accurately	Does not differentiate between different type of SCD, need to be validated	Electrical impedance of the sickle cells Detect Sickling RBCs
Imaging flow cytometry	Not reported	Not reported	Not reported	Robust test, can be automated to correlate the percentage of HbF and the percentage of sickled cells, biomarker of disease severity	Affected by agents that reduce polymerization of HbS, laborious	Used to quantify sickled cells
Optical tweezer to capture red blood	Not reported	Not reported	Not reported	Can be a monitor test, simple	Cannot indicate the severity of the disease in heterozygous states	Measuring red blood cell elasticity
Photoacoustic Flow cytometry	Not reported	Not reported	Not reported	Simple, low-cost, uses cell phone-like camera	It is not clear if it can be used to monitor the disease severity, cannot distinguish between sickle cell trait and sickle cell disease	Determine the RBCs Sickling
lateral flow Immunoassay sickle SCAN	90%	100%	98%	Simple, rapid	Relies on a polyclonal antibody, more expensive, low specificity and cross-reactivity, qualitative test, the intensity of	Identify HbC and HbS

					the band shows inconsistency, does not identify hemoglobin F, limit of detection of Hb A is 2%	
lateral flow Immunoassay HemoTypeSC	93.4%	99.9%	99.1%	Cost-effective, rapid, POC	Cannot detect all hemoglobin variants, does not differentiate between HbSS and sick-le-β0-thalassemia, misinterpretation of the result in cases with recent blood transfusion	HbAA, HbAS, HbAC, HbSC, and HbCC
Hem-eChip-Microelectrophoresis assay	100%	HbSS 98.7% Other type 100%	100%	Reliable. POC, inexpensive, simple	Interpretation requires skills, the need for web-based images for automated results	SCD-SS, SCD-SC, and SCD Trait Hb E Disease
SCD-AMPS 2-phase	90%	97%	77%	Inexpensive, simple POC	Interpretation is difficult, less reliable, affected by different condi tions that decrease the number of dense cells, may not be appropriate for	Identifies Hb S and Hb A
SCD-AMPS 3-phase	91%	88%	69%			Identifies Hb S, Hb A and Hb C

					neonatal screening, low sensitivity and specificity	
Paper-based hemoglobin solubility test	94.2%	97.7%	96.9%	Simple, rapid, inexpensive POC, does not need trained personal	Difficult to distinguish HbAS (trait) from HbSC, humidity can affect the test result, low sensitivity and specificity	Diagnosis of HbSS
Quartz crystal microbalance (QCM) sensor	Not reported	Not reported	Not reported	Reliable, simple, POC, low-cost	Not a diagnostic test	Determine RBC's elasticity
Electrochemical genosensor	1.23 × 105 ohmLm-mol-1 cm-2	Not reported	Not reported	Simple, low cost, POC	Determination of SCA trait only	Detect βAβS
Surface plasmon resonance-based biosensor	Not reported	Not reported	Not reported	Simple, rapid	Needs PCR product, needs to be validated	Detect βSβS
The Pyrosequencing technique (PyS)	98.2%	Not reported	Sickle cell anemia 98.7% sickle cell-hemoglobin C disease with 98.7%, and the heterozygous with 92.2%,	Diagnose heterozygous SCD, simple, fast, low cost, suitable for large-scale	Misclassification, and false negatives, depend on primer design	Detect βSβS, Sβ0 thalassemia, Sβ+ thalassemia, and sickle-hemoglobin C

Source: Arishi, W. A., Alhadrami, H. A., &Zourob, M., (2021). Techniques for the detection of sickle cell disease: a review. Micromachines, 12(5), 519. https://doi.org/10.3390/mi12050519:

1.4.5.1. Blood Flow

The life expectancy is 20–30 years less than that of normal adults; however, the advancement in therapy is prolonging the survival rate. In the US, the disease is diagnosed with prenatal screening which is based on Hb electrophoresis, amplifying the types of Hb and identifying various hemoglobinopathies. Along with a complete CBC test, it is checked with reticulocyte count, LDH level, complete metabolic panel, bilirubin level, and crossmatch for any transfusion therapy if possible (Sedrak & Kondamudi, 2023). Reticulocytes are the immature and young RBCs produced in the bone marrow. They are usually released in the blood only after the cells have attained maturity, and their numbers are fairly insignificant in healthy persons, which is about 0.45–1.8%. However, patients with sickle cell disease have large numbers of reticulocytes in the body as it is rapidly producing more RBCs to address anemia, and their reticulocyte count can be 2–3% or more (David et al., 2018).

The monitoring and diagnosis of this condition entail several approaches, and this list is increasing with the development of newer technology. Several approaches, as compiled in a review, have been presented below.

1.4.6. Hemoglobinopathies

These conditions are inherited disorders consisting of abnormal molecules of hemoglobin, as in the case of sickle cell disease, where the molecules are altered structurally, which impacts the normal function of RBCs. They are the most common inherited diseases around the globe but have recently risen in Central and Northern Europe, which could be due to immigration. The normal range of Hb found in tests is as shown below (Kohne, 2011) (Table 1.6) (Figure 1.8).

Table 1.6. The Normal Range of Hb Found in Tests

Haemoglobin F (Fetal Hemoglobin): This makes up 1%–2% of Hb found in adults having two alpha and two gamma protein chains. The fetus produces this primary Hb during pregnancy and its production gradually reduces after birth until 2 years of age.
Haemoglobin A: This makes up 95%–98% of Hb found in adults and is made of two alpha and two beta protein chains.
Haemoglobin A2:This makes up 2%–3% of Hb in adults and is made of two alpha and two delta protein chains.

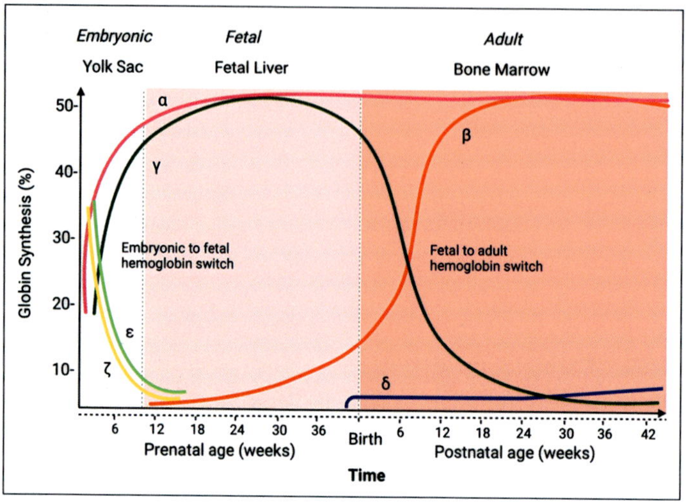

Figure 1.8. *The changes in how the globin gene expression documents a temporal change in expression.*

Source: Segura, E. E. R., Ayoub, P. G., Hart, K. L., & Kohn, D. B., (2023). Gene therapy for β-hemoglobinopathies: From discovery to clinical trials. Viruses, 15(3), 713. https://doi.org/10.3390/v15030713: © 2023 by the authors. Licensee MDPI, Basel, Switzerland. This article is an open access article distributed under the terms and conditions of the Creative Commons Attribution (CC BY) license (https://creativecommons.org/licenses/by/4.0/).

1.4.7. Thalassemia

This is observed as a group of genetic disorders that impact the production of hemoglobin. Based on the type of thalassemia, the red blood cells can be microcytic (smaller in size) and scarce owing to the insufficient production of alpha/beta globin chains. α and β-thalassemias are genetically heterogeneous diseases and the RBC transfusions entailed as part of the treatment classify them as transfusion-dependent thalassemia (TDT) and non-transfusion-dependent thalassemia (NTDT) (Figure 1.9).

Major thalassemia is identified by a lowered Hb level of <7 g/dl, mean corpuscular Hb (MCH) > 12 < 20 pg, and mean corpuscular volume (MCV) > 50 < 70 fl. Minor Thalassemia is identified by the lowered values of MCH and MCV coupled with elevated Hb A2 levels (Galanello & Origa, 2010). The major complications (the focus is on cardiovascular diseases) and their management as compiled by researchers et al. (2023) have been presented Table 1.7.

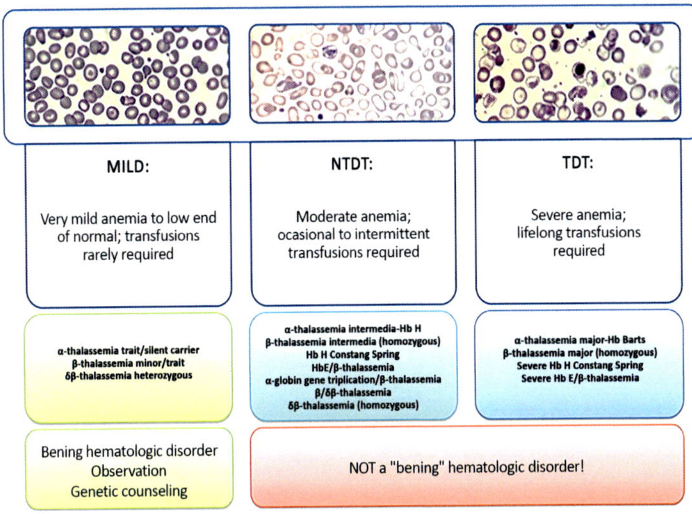

Figure 1.9.*Depicts the types of this condition.*

Source: Sanchez-Villalobos, M., Blanquer, M., Moraleda, J. M., Salido, E. J., & Perez-Oliva, A. B., (2022). New insights into pathophysiology of β-thalassemia. Front. Med., 9, 880752. doi: 10.3389/fmed.2022.880752: Copyright © 2022. This is an open-access article distributed under the terms of the Creative Commons Attribution License (CC BY). The use, distribution or reproduction in other forums is permitted, provided the original author(s) and the copyright owner(s) are credited and that the original publication in this journal is cited, in accordance with accepted academic practice. No use, distribution or reproduction is permitted which does not comply with these terms.

Table 1.7. The major complications and their management as compiled by Akiki et al. (2023)

Type of Complication	Causes and Risk Factors	Prevention and Management
Cardiomyopathy	• Iron overload • Chronic anemia • Reduced oxygen tissue delivery • Co-existent liver disease • Myocarditis • Pericarditis	• Iron chelation therapy. • Angiotensin II receptor blockers. • Diuretics: Spironolactone antagonists and digitalis glycosides. • Pacemaker • Heart transplant

Pulmonary Hypertension	• Hypercoagulability • Chronic hemolysis • Iron overload • Older age • Splenectomy • Hypoxia • Anemia • Female > Male • Hepatitis C	• If thromboembolism etiology, start anticoagulation • Blood transfusion • Iron Chelation • Sildenafil • Bosetan • Epoprostenol • Hydroxyurea
Arrhythmias	• Iron overload • Atrial Dilatation • Anemia • Hemolysis • Inflammation • Fibrosis and remodeling	• Iron Chelation therapy • Antioxidants • Rate and Rhythm control therapy • Ablation (If needed) • Anticoagulation therapy (Warfarin or DOACS)
Vasculopathies	• Hemolysis • Iron overload • High serum ferritin level	• Target hemolysis • Erythrocyte decompartmentalization • Arginase • Iron chelation
Sudden Cardiac Death	• Delayed cardiac disease diagnosis • Iron overload • Long QTd and JTd on ECG • Male > Female	• Careful ECG monitoring • 24-h Holter in high-risk patients.

Source: Akiki, N., Hodroj, M. H., Bou-Fakhredin, R., Matli, K., & Taher, A. T., (2023). Cardiovascular complications in β-thalassemia: getting to the heart of it. Thalassemia Reports, 13(1), 38–50. https://doi.org/10.3390/thalassrep13010005:

1.4.8. Chronic Kidney Disease (CKD)

Erythropoietin (EPO) is a hormone produced by the kidneys. In CKD, there is a significantly lowered production of this hormone which stimulates the production of RBCs in the bone marrow resulting in anemia (Figure 1.10).

The mean RBC count in elderly CKD patients is about 3.29 ± 0.79 × 106/μl. A remarkable decline in this blood count was observed as the stages of CKD progressed. 60% of stage 5 CKD and 74% of stage 4 CKD patients showed levels of hemoglobin to be less than 10g/dl. Around 59% of the patients were anemic and as per the serum iron profile, it was a result of the chronic nature of the disease. Anemia is the foremost cause of concern in CKD patients and it keeps deteriorating as the condition progresses. Hence, timely checkup and evaluation of RBC and Hb levels aids in classifying the stage of anemia and helps in changing the treatment to prolong the life of such patients (Shastry&Belurkar, 2019).

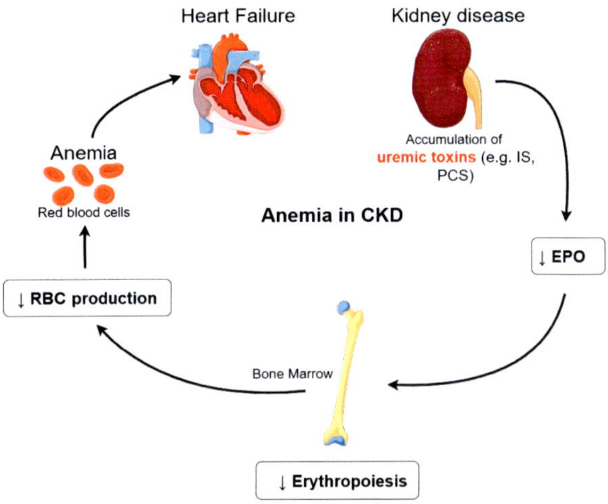

Figure 1.10. *The link between anemia and CKD is depicted.*

Source: Hamza, E., Metzinger, L., &Metzinger-Le Meuth, V., (2020). uremic toxins affect erythropoiesis during the course of chronic kidney disease: A review. Cells, 9(9):2039. https://doi.org/10.3390/cells9092039: © 2020 by the authors. Licensee MDPI, Basel, Switzerland. This article is an open access article distributed under the terms and conditions of the Creative Commons Attribution (CC BY) license (http://creativecommons.org/licenses/by/4.0/).

1.4.9. Thrombotic Microangiopathies (TMA)

Though rare, health conditions like HUS (hemolytic-uremic syndrome) and TTP (thrombocytopenic purpura), which are types of TMA, cause damage to the RBCs and fragmentation resulting in the formation of fragmented RBCs, also called schistocytes, in the blood. There is the formation of small blood clots in smaller blood vessels, which can result in damaging the organs, especially the brain and kidneys. The RBC count in TMA varies depending on whether it is HUS or TTP. A schematic of atypical HUS is depicted in Figure 1.11.

Studies show in HUS, hemolysis takes place lowering the RBC count, hematocrit levels, and Hb count and is mostly associated with anemia and its respective symptoms along with thrombocytopenia and acute damage to the kidneys. The enzyme ADAMTS13 prevents the formation of blood clots and is deficient in TTP causing uncontrolled clotting in the small blood vessels, resulting in anemia.

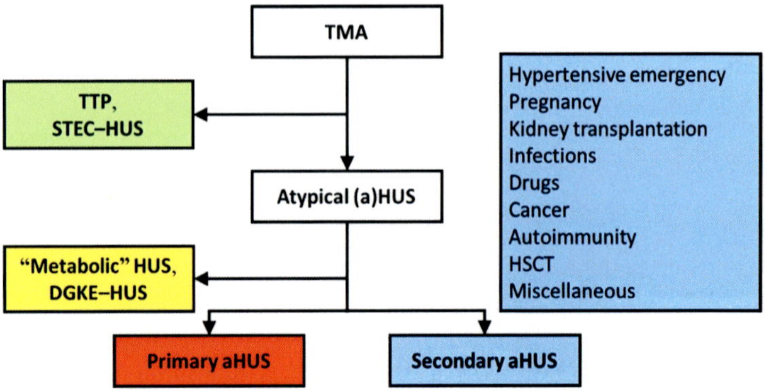

Figure 1.11. *A schematic of atypical HUS.*

Source: Timmermans SAMEG, & Van Paassen P., (2021). The syndromes of thrombotic microangiopathy: A critical appraisal on complement dysregulation. Journal of Clinical Medicine, 10(14), 3034. https://doi.org/10.3390/jcm10143034: © 2021 by the authors. Licensee MDPI, Basel, Switzerland. This article is an open access article distributed under the terms and conditions of the Creative Commons Attribution (CC BY) license (https://creativecommons.org/licenses/by/4.0/).

In both types, the RBC count is lowered significantly due to hemolysis and anemia. Further lab tests show increased levels in the markers of RBC breakdown – unconjugated bilirubin, low haptoglobin, and lactate dehydrogenase (LDH). If left untreated, the mortality rate is as high as 90%; however, with exchange in plasma, this can be lowered to 20%. Exchange of plasma with fresh samples from donors is known to be the first line of protection administered rapidly to save the patients (Arnold et al., 2017).

1.4.10. Myelodysplastic Syndromes (MDS)

These syndromes are a group of disorders where the bone marrow fails to make sufficient quantities of healthy blood cells. MDS shows altered morphology in RBCs. Clinically, MDS is expressed as a clonal disorder of hematopoietic stem cells resulting in ineffective hematopoiesis and dysplasia in the bone marrow, and in some patients, it is transformed into acute myeloid leukemia (AML). The origin of this condition has been comprehensively presented by Fontenay et al. (2021) in the Figure 1.12.

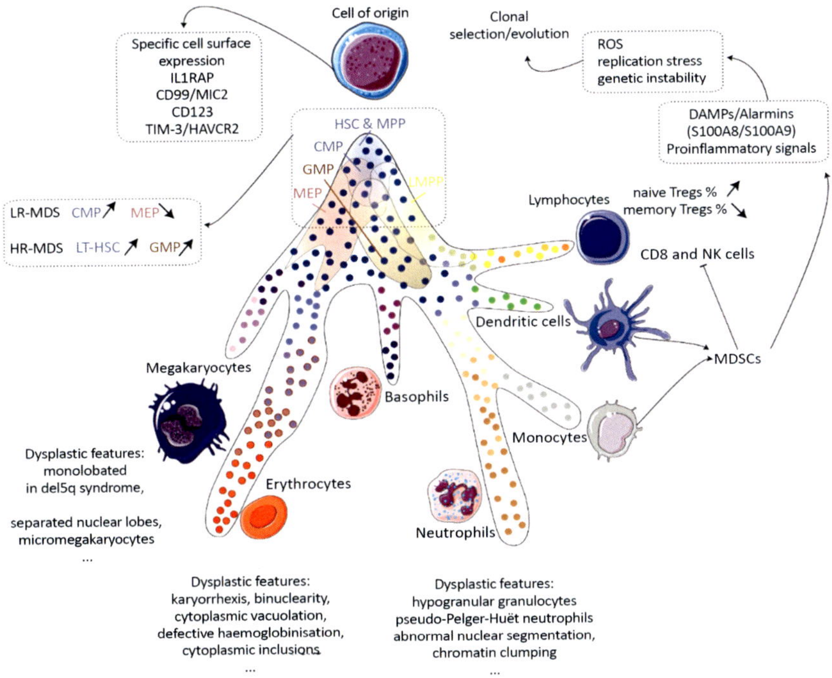

Figure 1.12. An account of myelodysplastic syndromes (MDS)

Key: "myeloid-derived suppressive cells (MDSCs), IL1RAP: IL-1 receptor accessory protein; TIM-3/HAVCR2: T-cell immunoglobulin mucin-3/Hepatitis A virus cellular receptor 2; LR-MDS: low-risk MDS; HR-MDS: high-risk MDS; HSC: hematopoietic stem cell; LT-HSC: long-term hematopoietic stem cell; MPP: multipotent progenitors; CMP: common myeloid progenitor; GMP: granulocytic myeloid progenitor; MEP: megakaryocytic-erythroid progenitor; Treg: regulator T cells; NK: natural killer; ROS: reactive oxygen species; DAMPs: damage-associated molecular patterns"

Source: Fontenay M, Farhat B, Boussaid I., (2021). Pathophysiology of myelodysplastic syndromes. Hemato. 2021; 2(3):477–495. https://doi.org/10.3390/hemato2030030: © 2021 by the authors. Licensee MDPI, Basel, Switzerland. This article is an open access article distributed under the terms and conditions of the Creative Commons Attribution (CC BY) license (https://creativecommons.org/licenses/by/4.0/).

It is essential to record that certain changes in RBCs vary based on the health condition/disease, its severity and individual patient health responses. The diagnosis usually consists of a physical examination, clinical history, laboratory tests like CBC, hemoglobin electrophoresis and peripheral blood smear and at times some more specialized tests. Altered and lower RBC count is a feature of MDS that differs based on the subtype and severity in patients. The patients are anemic and since the bones fail to produce healthy RBCs, the Hb levels are considerably low. In some cases, the RBC count may be normal but since the RBCs are abnormal based on their shape, size, and maturity, the condition persists. Treatment may include blood transfusion to treat the anemia, bone marrow transplants in some cases, and medication for reviving RBC production (Dotson &Lebowicz, 2022).

1.5. CHANGES IN WBC COUNT

Changes in the count of WBCs are usually indicators of severe diseases. They play an essential role in the immune response of the body. Any alteration in their count supplies essential information regarding the presence and nature of inflammation, infections, or any health condition diagnosed on detailed study. A higher WBC count indicates bacterial, viral, or parasitic infections, an allergic reaction, inflammatory diseases, leukemia, or tissue damage post-surgery or burn. A lower WBC count indicates damage in the bone marrow, cancers affecting the bone marrow, autoimmune disorders, and HIV/AIDS. Below are some examples of how changes in the values of WBCs help in the diagnosis of these conditions (Kabat et al., 2017).

1. **Infections:** *Bacterial Infections:* WBC count increases as a result of response to bacterial infections. It is mainly due to the neutrophils which are involved in attacking bacterial pathogens. This condition is called neutrophilia. Bacterial infections are generally easier to treat compared to viral and parasitic infections.

i. **Viral Infections:** Some viral infections can lead to a decrease in WBC count or in neutrophils, causing lymphocytosis. Lymphocytes become more prominent in WBC during viral attacks.

ii. **Fungal and Parasitic Infections:** These infections lead to changes in WBC count due to the immune response triggered by the microbes/pathogens (Kabat et al., 2017).

2. **Allergic Reactions:**Allergic reactions result in an increase in eosinophil levels, which is a response to parasites and allergic conditions.

3. **Inflammatory Conditions:***Autoimmune Disorders:* Health conditions such as lupus, inflammatory bowel syndrome, and rheumatoid arthritis can cause an increase in WBC count due to chronic inflammation.

Figure 1.13 illustrates the significance of the CBC.

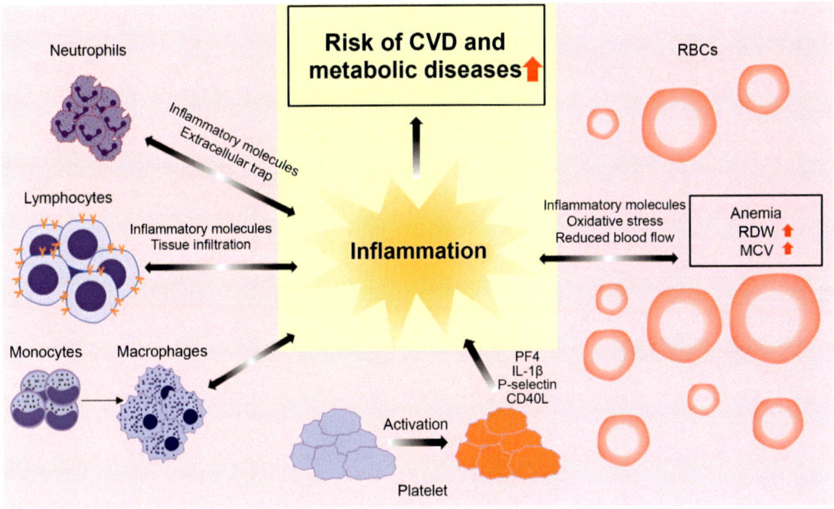

Figure 1.13. *The significance of the CBC.*

Source: Seo I-H, Lee Y-J., (2022). Usefulness of complete blood count (CBC) to assess cardiovascular and metabolic diseases in clinical settings: A comprehensive literature review. Biomedicines. 2022; 10(11), 2697. https://doi.org/10.3390/biomedicines10112697: © 2022 by the authors. Licensee MDPI, Basel, Switzerland. This article is an open access article distributed under the terms and conditions of the Creative Commons Attribution (CC BY) license (https://creativecommons.org/licenses/by/4.0/).

4. **Cancers:**

i. **Leukemia:** This cancer has its origin in the bone marrow and it causes an overproduction of abnormal WBCs, thus increasing their count. This cancer can be detected based on the high WBC count with irregular distribution of the different types of WBCs (Tigner et al., 2022).

5. **Stress and Trauma:** Circumstances like trauma, surgery, or extreme physical stress result in an increase in the WBC count. This is generally the body's response to tissue damage and inflammation.

6. **Bone Marrow Disorders:**

   – **Myelodysplastic Syndromes (MDS):** In these groups of disorders, the bone marrow fails to produce enough quantity of healthy blood cells. This results in an unusually high or low WBC count depending on the specific types of MDS.

It is necessary to take notice of the fact that though the WBC count acts as an essential tool for diagnostic information, it is just one part of the entire puzzle. A thorough and complete clinical examination inclusive of the patient's physical health, medical history, and any other lab tests are essential for correct diagnosis and management of the health condition. Also, the WBC count varies based on different labs and the techniques used by them, the patient's gender, age, and current health status (Mohammad, 2018).

7. **Cardiovascular Disorders:**The magnitude and regulation of the correspondence linked between WBC and coronary heart disease (CHD) is uncertain owing to the sensitivity of reverse causation. Zhao and team reported that larger WBC and neutrophil counts are linked to higher risks of CHD. The current sensitive diagnostic tests to analyze cardiovascular health include ultrasounds, electrocardiography, cardiac damage markers, and angiography which are expensive and time-consuming. Studies in the future to diagnose CHD through WBC would be a novel invention (Zhao et al., 2023).

1.6. CHANGES IN PLATELET COUNT

Platelet count alterations are most commonly noticed in several diseases and health conditions. Though tiny, they play a major role in the clotting of blood and healing of wounds. Changes in platelet count signify conditions mentioned below.

1. **High Platelet Count (Thrombocytosis):** This condition is marked by an unusually high platelet count. It is further classified as Primary and Secondary Thrombocytosis.

i. **Primary Thrombocytosis:** This condition is a result of the unusual overproduction of platelets in the bone marrow leading to complications occurring in blood clotting.

A graphical representation of the pathophysiology of immune thrombocytopenia (ITP) by Singh et al. (2021) has been given Figure 1.14.

ii. **Secondary Thrombocytosis:** This condition is a result of the reactive response to underlying health conditions like inflammation, infections, removal of the spleen, or cancers.

2. **Low Platelet Count (Thrombocytopenia):** This condition is marked by an unusually lower than normal platelet count which results in a higher risk of bleeding and a strenuous impact on clotting (Bleeker and Hogan, 2011).

i. **Immune Thrombocytopenic Purpura (ITP):** This is an autoimmune disorder where the person's immune system attacks self-cells while destroying the platelets, causing a low platelet count.

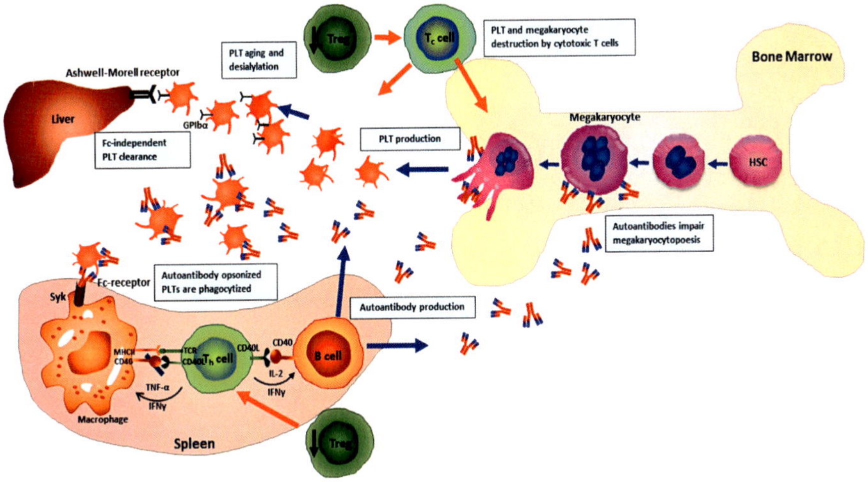

Figure 1.14.*A graphical representation of the pathophysiology of immune thrombocytopenia (ITP).*

Source: Singh A, Uzun G, &Bakchoul T., (2021). Primary immune thrombocytopenia: novel insights into pathophysiology and disease management. Journal of Clinical Medicine, 10(4), 789. https://doi.org/10.3390/jcm10040789: © 2021 by the authors. Licensee MDPI, Basel, Switzerland. This article is an open access article distributed under the terms and conditions of the Creative Commons Attribution (CC BY) license (http://creativecommons.org/licenses/by/4.0/).

ii. **Heparin-Induced Thrombocytopenia (HIT):** This condition is a result of an unfavorable response to Heparin, which is a blood thinner. This results in a remarkable decline in the platelet count, causing an increase in the risk of blood clotting (Ahmed et al., 2007).

iii. **Bone Marrow Disorders:**Conditions such as myelodysplastic syndromes, leukemia, and aplastic anemia can impact the production of platelets, resulting in low platelet counts (Luu et al., 2018).

3. **Dengue Fever:** This viral infection causes a remarkable decrease in the platelet count, resulting in thrombocytopenia, and in severe cases, it can cause complications in clot formations (Das et al., 2022).

Several viruses can activate platelets as summarized in the illustration Figure 1.15 by Ostermeier et al. (2022): Human immunodeficiency virus-1 (HIV-1), dengue virus (DENV), influenza A virus (IAV), and severe acute respiratory syndrome coronavirus 2 (SARS-CoV-2) bind to receptors on the plasma membrane of platelets. Examples of receptors include CXCR/CCR receptors, DC-SIGN, CLEC 2, FCγRIIα, ACE2, and TLRs 3/4/7/8. Subsequent activation leads to the release of α-granule release, the release of soluble CD40L (sCD40L), and the release of extracellular vesicles (EV) and microvesicles (MV).

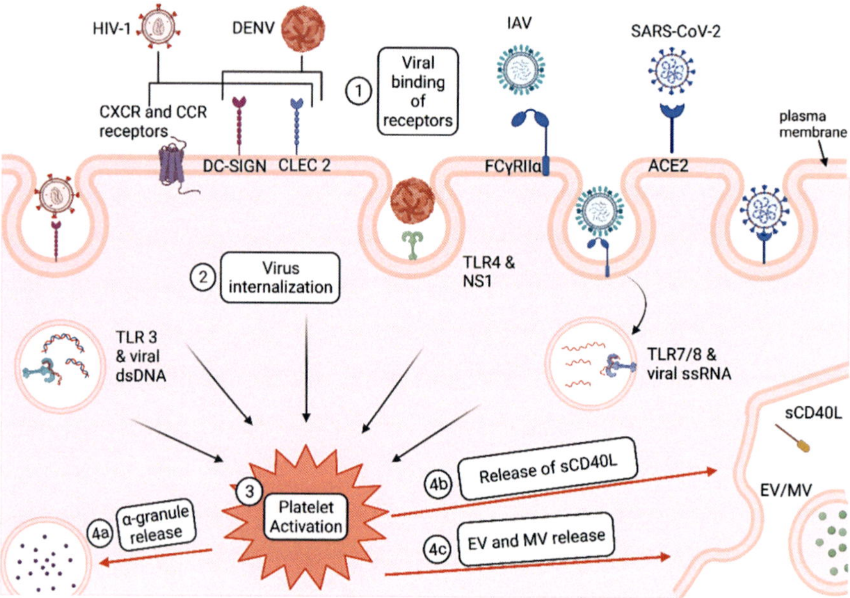

Figure 1.15. Activation of platelets by a few viruses

Source: Ostermeier B, Soriano-Sarabia N, &Maggirwar SB., (2022). Platelet-released factors: Their role in viral disease and applications for extracellular vesicle (EV) therapy. International Journal ofMolecularSciences, 23(4), 2321. https://doi.org/10.3390/ijms23042321: © 2022 by the authors. Licensee MDPI, Basel, Switzerland. This article is an open access article distributed under the terms and conditions of the Creative Commons Attribution (CC BY) license (https://creativecommons.org/licenses/by/4.0/).

4. **HUS (Hemolytic Uremic Syndrome):** This condition is marked by the annihilation of RBCs and platelets. This usually happens after an infection and can cause issues with the kidneys and low platelet counts (Karpman et al., 2006).
5. **Cancer:** Certain cancers can cause variation in the platelet count, for example, in patients with cancers involving bone marrow can encounter thrombocytopenia (Giannakeas et al., 2022).

6. **Radiation and Chemotherapy:** Cancer treatments like radiation and chemotherapy suppress the activity of the bone marrow which in turn impacts the production of platelets resulting in thrombocytopenia (Giannakeas et al., 2022).

Cancer cells and platelets have a documented complex and bi-directional association. Yu et al. (2021) have extensively summarized this association as shown in Figure 1.16.

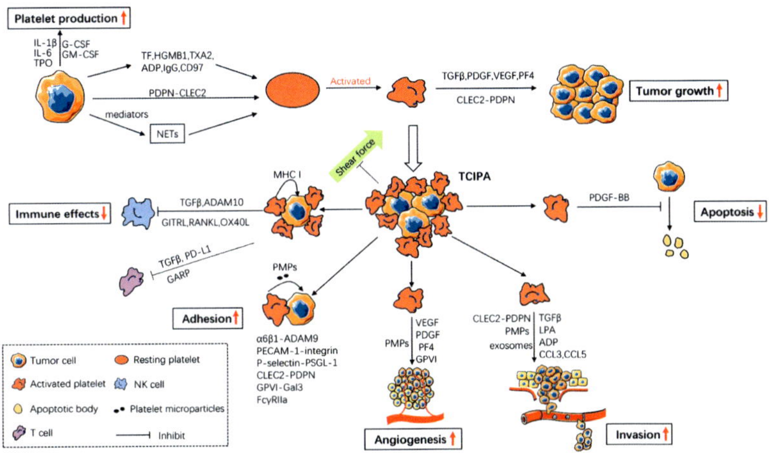

Figure 1.16. The connections between cancer cells and platelets

Key: "IL-1β, Interleukin-1β; IL-6, Interleukin-6; TPO, thrombopoietin; G-CSF, Granulocyte- Colony-Stimulating Factor; GM-CSF, Granulocyte-Macrophage-Colony-Stimulating Factor; TF, Tissue Factor; HGMB1, High-Mobility Group Box1; TXA2, Thromboxane A2; IgG, immunoglobulin G; PDPN, Podoplanin; CLEC2, C-type lectin receptor type 2; NETs, Neutrophil Extracellular Traps; TGFβ, Transforming Growth Factor β; PDGF, Platelet-Derived Growth Factor; VEGF, Vascular Endothelial Growth Factor; PF4, Platelet Factor 4; TCIPA, Tumor Cell-Induced Platelet Aggregation; MHC I, MHC class I; ADAM10, Disintegrin And Metalloproteinase Domain-Containing Protein 10; GITRL, Glucocorticoid-Induced TNF Receptor-Related Ligand; RANKL, Receptor Activator of NF-κB Ligand; OX40L, Oxford 40 Ligand; GARP, Glycoprotein A Repetitions Predominant; PD-L1, Programmed Cell Death-Ligand 1; PMPs, Platelet Microparticles; PECAM-1, Platelet-Endothelial Cell Adhesion Molecule-1; PSGL-1, P-selectin Glycoprotein Ligand-1"

Source: Yu L, Guo Y, Chang Z, Zhang D, Zhang S, Pei H, Pang J, Zhao ZJ & Chen Y, (2021). Bidirectional interaction between cancer cells and platelets provides potential strategies for cancer therapies. Front. Oncol., 11, 764119. doi: 10.3389/fonc.2021.764119: Copyright © 2021. This is an open-access article distributed

Table 1.8. Neurodegenerative conditions and platelet abnormalities-please note that the references in numbers are cited by the authors who compiled the table-interested readers may read Leiter and Walker (2020)

Condition	Implication of platelets	Species/model	Reference
Alzheimer's disease	Increased platelet activation	Human	(35)
	Increased platelet β-secretase activity	Human	(36, 37)
	Platelet hyperactivity	APP23 mice	(38)
	Increased adhesion to subendothelial matrix components	3xTg-AD mice	(39)
	Platelet inclusions in cerebral blood vessels	APP_SweDI mice	(40)
	Platelets enhance formation of amyloid-β aggregates in cerebral vessels	APP23 mice	(41)
	Platelets promote neuroinflammation and vessel damage	APP_SweDI mice	(42)
	APP and amyloid-β influence platelet function	Human/APP-KO, C57BL/6 and APP23 mice	(41, 43–51)
Huntington's disease	Increased platelet mHtt protein levels	Human	(52)
	Increased platelet aspartate and glycine levels	Human	(53, 54)
	Platelets promote blood brain barrier permeability	Human	(52)
	Impaired platelet adenosine A receptor signaling	Human	(55, 56)
	Impaired platelet nitric oxide metabolism	Human	(57)
	Elevated platelet mitochondrial monoamine oxidase activity	Human	(58, 59)
Parkinson's disease	Increased mean platelet volume	Human	(60)
	Decreased platelet glutamate uptake	Human	(27)
	Reduction in vesicular monoamine transporter 2 mRNA	Human	(61)
	Platelet mitochondrial dysfunction	Human/ Cybrid model	(62–66)
Amyotrophic lateral sclerosis	Increased platelet TDP-43 levels	Human	(67)
	Reduced complex IV activity in platelet mitochondria	Human	(68)
	Altered platelet mitochondrial membrane potential	Human	(69)
	Altered platelet mitochondrial morphology	Human	(70)
	Altered platelet activation and morphology	Human	(70)
	Enlarged mitochondria, degenerating mitochondrial vacuoles and neurofilament aggregations	Cybrid model	(71–73)
	Decreased platelet serotonin levels	Human	(74)
Multiple sclerosis	Increased platelet activation	Human	(75)
	Platelets drive neuroinflammation in the spinal cord	EAE mice	(76, 77)
	Platelet-neuron associations are associated with neuroinflammation in the hippocampus	EAE mice	(14)
	Altered serotonin release from dense granules	Human/EAE mice	(78)
Prion diseases	Platelets carry infectious prions	Deer	(79)
	Platelets are capable of transmitting disease phenotypes	Deer and sheep	(79, 80)

Source: Leiter O, & Walker TL, (2020). Platelets in neurodegenerative conditions—friend or foe? Front. Immunol., 11:747. doi: 10.3389/fimmu.2020.00747:

cited, in accordance with accepted academic practice. No use, distribution or reproduction is permitted which does not comply with these terms.

Note from the Author: It must be noted that the changes in platelet counts are just one feature in a vast clinical study. These values are used in conjunction with patient history and other diagnostic tools for the identification and treatment of diseases and health conditions of patients. It is advised to not rely on self-research on the internet but consulting a healthcare provider is a must for proper and rapid assessment and counsel for any symptoms.

7. **Liver Diseases:** Advanced and chronic liver diseases can cause variations in the production and functions of platelets. This can develop in conditions like cirrhosis.
8. **Chronic Inflammatory Disorders:** Conditions like SLE (systemic lupus erythematosus) and rheumatoid arthritis can impact the platelet count due to the inflammatory response triggered by these disorders (Seyoum et al., 2018).

Interestingly, besides the role of platelets in hematology, research has revealed that neurodegenerative conditions also exhibit platelet dysfunction. For example, Leiter and Walker (2020) summarized neurodegenerative conditions and platelet abnormalities. Their findings are presented Table 1.8.

1.7. BLOOD CELLS AND COVID-19

"Let us brace our offended solitary mindset in spite of COVID-19 as the world's finest hour perpetual uncertainty, let us brace our hearts as the very center of our hearts is an armored supreme excellence to be victorious by an unequivocal perseverance and hope, let us brace unconditional love, kindness, and triumph to be the masters of our own chain of destiny, let us brace unity and oneness to beckon for victorious war against fear and the inherent vice of crisis at its peril."

—Dr. Tony Beizaee

COVID-19, caused by the coronavirus SARS-CoV-2, led to the recent worldwide pandemic and it certainly shows several changes in the CBC test. It is to be noted that the severity of the disease varies in infected people based on their immune response and their lifestyle. The patients can be classified as severe, moderate, and mild. Hematological variables like WBC count, Hb, platelets, neutrophils, and lymphocytes are involved in the severity of the infection. Patients encountered a broad array of diseases ranging from mild illness to varied degrees of severity of pneumonia to ARDS (acute respiratory distress syndrome) along with multiple organ failures, sepsis, and death.

Table 1.9. Various effects of COVID-19, including the hematological effects

Systems	Main Diagnosis	Features	Possible Mechanisms	Prognosis	References
Respiratory System	Acute respiratory distress syndrome (ARDS)	Extensive bilateral diffuse alveolar damage with cellular fibromyxoid exudates; desquamation of pneumocytes and hyaline membrane formations; diffusion impairment and pulmonary fibrosis.	SARS-CoV-2 spike S1 domain protein binding to ACE2 receptor; Post Acute Respiratory Distress Syndrome fibrosis with diffuse alveolar damage	Pulmonary function deficit 6 months after infection; extensive diffuse impairment; Long-term in-situ thrombosis	David, et al., (2018); Köhne, (2011); Galanello, &Origa, (2010); Shastry, &Belurkar, (2019); Arnold, Patriquin, &Nazy, (2017); Dotson, &Lebowicz, (2022); Kabat, Kim, Manson, Lessin, Lin, Wassertheil-Smoller, & Rohan, (2017); Tigner, Ibrahim, & Murray, (2022); Mohammad, (2018); Zhao, Liu, Chen, Yang, Dong, & Bai, (2023); Bleeker, & Hogan, (2011)
Cardiovascular System	Endothelitis; Micro-thrombosis, Capillary damage; hypercoagulability; microangiopathy; thromboembolism; myocarditis; atrial fibrillation; supraventricular tachycardia	Increased target-to-blood pool ratio; capillary disturbance; impaired oxygen diffusion.	Cytokine storm and macrophage activating syndrome-caused endothelial dysfunction.	Majority (81%) of the COVID-19 myocarditis patients survived the acute episode; ongoing subclinical myocarditis may evolve into myocardial dysfunction and sudden cardiac death.	Ahmed, Majeed, & Powell, (2007); Luu, Gardiner, & Andrews, (2018); Das, Abreu, Harris, Shrader, &Sarvepalli, (2022); Karpman, Manea, Vaziri-Sani, Ståhl, &Kristoffersson, (2006); Giannakeas, Kotsopoulos, Brooks, Cheung, Rosella, Lipscombe, &Narod, (2022)

Haematological System	Thromboembolism	Elevated convalescent D-dimer and C-reactive protein levels; persistently increased biomarkers of inflammation.	N/A	Prognostic biomarkers for monitoring clinical progression of Long COVID-19 patients need to be investigated	Al-Saadi&Abdulnabi, (2022); Palladino, (2021); Wu, & McGoogan, (2020)
Urinary System	Acute kidney injury; renal failure;	Declined glomerular filtration rate (eGFR); kidney infarction	High abundance pf ACE2 expression in kidneys.	Significant risks of mortality and morbidity	Lippi, Plebani, & Henry, (2020); Seyoum, Enawgaw, &Melku, (2018); Cossarizza, De Biasi, et al., (2020); Camp, & Jonsson, (2017)
Digestive System	Gastrointestinal impairment and dysfunction; hepatic and cholestatic liver injury; pancreatic injury	Bowel diffuse damage; Enterocytes desquamation, edema, small bowel dilation, lymphocytes infiltration and mesenteric nodes hemorrhage and necrosis.	Rich in ACE2 and furin expression; fecal-oral transmission; plasma cells and lymphocytic infiltrations into lamina propria of intestinal tissues.	The liver enzymes remained persistently elevated 14 days after discharge, while the liver functions in majority survivors normalized 2 months after hospital discharge	Yamada, Wakabayashi, Yamaji, Chopra, Mikami, Miyashita, & Murakami, (2020); Zhang, et al., (2020); Soraya, &Ulhaq, (2020); Zini, Bellesi, Ramundo, & D'Onofrio, (2020); Nazarullah, Liang, Villarreal, Higgins, &Mais, (2020); Zhang, Zhou, Zhu, Song, Feng, Qiu, & Zhang, (2020); Lu & Wang, (2020)

Neurological System	Mood changes; cognitive difficulties; headache; fatigue; dizziness; memory loss; confusion; and attention deficit.	Hypoxic injury; microbleedings; neuronal inflammations.	Blood vessel damage, impaired oxygen supply, viral infiltration into the central nervous system and inflammatory cytokines-mediated cellular damage; indirect T-cell and microglia damage in the brain, similar to strokes and neuroinflammatory diseases.	Over 40% survivors without prior psychiatric conditions lived with depression within 90 days of recovery from severe COVID-19 associated respiratory failure, while 70% of them did not receive treatment for depression	Bastug, Bodur, Erdogan, Gokcinar, Kazancioglu, Kosovali, &Duman, (2020); Ahbap, Sakaci, Kara, Sahutoglu, Koc, Basturk, &Kayalar, (2016); Zhang, Guo, Lei, Liu, Wang, Wang, & Cao, (2021); Zhou, Fu, Zheng, Wang, Zhao, Qi, & Wang, (2020); Zhang, Dong, Cao, Yuan, Yang, Yan, &Xie, (2020); Mu, Yi, Wang, Wang, Zhang, Chen, & Wang, (2021)
Metabolic System	Hyperglycemia without diabetic mellitus; new-onset diabetic mellitus; starvation ketoacidosis	High blood glucose level; impaired glucose metabolism	Intruding pancreatic β-islet cells, triggering autoimmune responses because of the exposure of the antigen from damaged islet cells.	Long-term treatment of diabetic mellitus is needed.	Rodriguez, Pekkarinen, Lakshmikanth, Tan, Consiglio, Pou, & Bjorkman, (2020); Qin, Zhou, Hu, Zhang, Yang, Tao, & Jiang, (2020); Tan, Zhou, Zhou, Hu, & Long, (2021); Thomas, Stefanoni, Dzieciatkowska, Issaian, Nemkov, Hill, & Hansen, (2020)

Source: Yan Z, Yang M, & Lai C-L., (2021). Long COVID-19 syndrome: A comprehensive review of its effect on various organ systems and recommendation on rehabilitation plans. Biomedicines, 9(8):966. https://doi.org/10.3390/biomedicines9080966: © 2021 by the authors. Licensee MDPI, Basel, Switzerland. This article is an open access article distributed under the terms and conditions of the Creative Commons Attribution (CC BY) license (https://creativecommons.org/licenses/by/4.0/).

Most patients developed pneumonia in around 20–30% of the patients resulting in respiratory failure, and the patients needed to be intubated and put on ventilator support. Globally, RT-PCR (reverse transcription polymerase chain reaction) is the most trusted test for detecting the disease (Palladino, 2021).

COVID-19 clinical manifestations range from asymptomatic to acute to severe lung damage caused by the virus entry, its load, and the host's immunity. Some clinical presentations are as shown in the table below. Gastrointestinal and respiratory infections are linked with the majority of COVID-19 cases (Al-Saadi et al., 2021). Lab findings showed a decrease in lymphocyte and WBC count. Neutrophilia, increased CRP (C-reactive protein), abnormal PCT (procalcitonin), and erythrocyte sedimentation rate were noted in most patients. Alterations in the blood count were indicative of the severity and stage of the disease (Wu & McGoogan, 2020). An evidence-based summary by Yan, Yang, Lai (2021) documenting the various effects of COVID-19, including the hematological effects, is shown Table 1.9.

1.8. SOME COMMON CBC FINDINGS IN RELATION TO THE INFECTION

1.8.1. Platelets

Platelet count is an essential variable that aids in evaluating the severity of diseases, for example, multiorgan dysfunction syndrome. Thrombocytopenia also indicates the severity of the disease and rates of coagulation, as it was noted to have increased significantly in the later stages of the disease (Wu & McGoogan, 2020). Studies have established a connection between a low platelet count and increased severity and higher mortality (Lippi et al., 2020).

The pathway adapted by the coronavirus to interfere with the blood cycle is still unknown. Scientists assume three probabilities in this context:

a. Destruction of the platelets by the immune system;
b. Virus directly infects the bone marrow cells, resulting in hindrance in the synthesis of platelets; and
c. Clustering of platelets in the lungs, forming microthrombi and eventually consuming more platelets (Wu & McGoogan, 2020).

Studies by Seyoum and team show the association of viruses with megakaryocytes resulting in a considerable decrease in the synthesis of platelets (Seyoum et al., 2018). Inflammation coupled with viral infection causes damage to the pulmonary capillary system. Here, the pulmonary endothelial cells and damaged lung tissues can result in the rupturing of megakaryocytes, in turn elevating the consumption of platelets. Studies also show that the virus can trigger the immune mechanisms and autoantibodies, reducing the platelet count. The infection results in damage to the lung endothelial cells and lung tissues that can trigger platelets forming microthrombi, ultimately causing more consumption of platelets. Hypoxemia coupled with the platelet count is also used to determine the severity of the infection with 96% accuracy. Compared to severely ill non-COVID-19 patients, the COVID-19 patients had remarkably higher MPV (mean platelet volume) along with an increase in the IPF (immature platelet fraction). COVID-19 patients showed IPF values of 7.5 x 10^9/L or more compared to non-COVID-19 patients with IPF of ≥ 8%, indicating an escalated production of large and immature platelet cells (Wu & McGoogan, 2020).

1.8.2. White Blood Cells

Lymphocytes: Lymphopenia (shortage of lymphocytes, < 1.0 x 10^9/L) is observed in 85% of severely Covid-19 infected patients. SARS-CoV-2 was compared to influenza, which is the most widely studied viral infection. The lymphocyte count was remarkably low in severe cases of the disease. Common causes included symptoms of fever, pneumonia, and respiratory tract symptoms, while the severe cases showed one of these three clinical expressions:

a. Respiratory rate >30 breaths/min indicative of shortness of breath;
b. Partial pressure of oxygen and oxygen concentration ≤ 300 mm Hg; and
c. Mean oxygen saturation ≤ 93% while in the resting phase

Advancement of up to 50% more lesions in 24 to 48 hours was observed on pulmonary imaging that was noted in severe cases. Severe cases also showed a reduction in the lymphocyte percentage rate compared to common COVID-19 and healthy control groups.

Scientists assume the CD8+ T cells to be associated with the infection. Pathways in response to IFN (interferon)-I and II were found to be linked to influenza-specific clusters during the inspection of disease-specific transcriptional signature, CD8+ T cells. Pathways of interleukin-1β and TNF (tumor necrosis factor) were significantly evident in COVID-19 clusters. All these suggest that COVID-19 can result in alterations in the peripheral blood parameters that need to be accounted for in the early stage of the disease to prevent its succession (Wu & McGoogan, 2020).

Lymphopenia can be a result of the emission of circulating lymphocytes in the inflamed tissues of the lungs or by direct attachment of the virus. Studies showed CD8+ T cells, CD4+ T cells, NK (natural killer) cells and accumulative lymphocyte cell counts reduced remarkably in patients with higher severity of COVID-19 and significantly lower in ICU patients. Also, the lowest number of T cells was noted in patients more than 60 years of age which can be suggestive of increased risk in elderly patients.

It was found that T-cells were depleted as the severity of COVID-19 progressed in the patients. This was a result of the triggering of expression by some immune-inhibitory molecules during the course of infection. It is known that the continuous triggering by the Coronavirus will result in the depletion of T-cells causing a decline in cytokine production and ultimately resulting in cellular dysfunction. When treated with chloroquine, there was an increase in the number of NK cells and T cells. Hence, it was hypothesized that in patients infected with Coronavirus exhibiting severe inflammation in the lungs, the virus corresponds to the depletion of cytotoxic lymphocytes in the initial phase which may lead to the disease (Palladino, 2021). Studies by Cossarizza and team showed a low value of T-cells, a decline in memory T-cells, and an increase in the naive helper T-cells in patients affected severely by COVID-19 (Cossarizza et al., 2020). The most important factor for recovery from the disease is an increase in the lymphocyte count.

In order to illustrate these findings, sample blood smears of hematological abnormalities from COVID-19 patients requiring intensive care unit (ICU) admission assayed and published by Lin et al (2022) are shown in Figure 1.17.

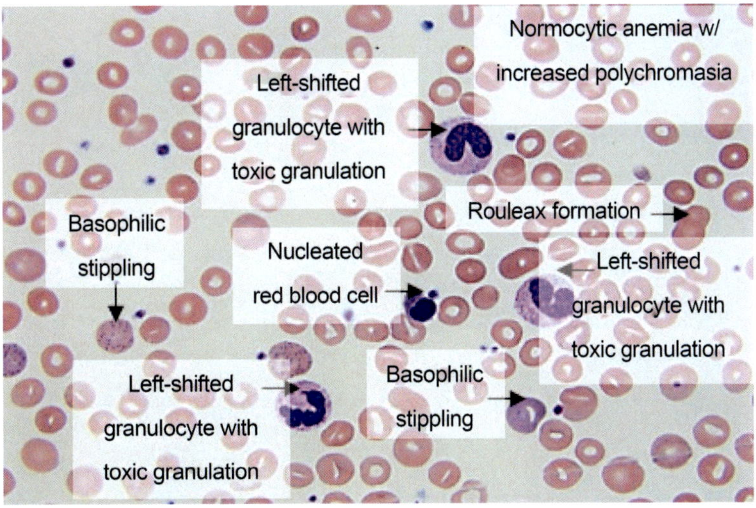

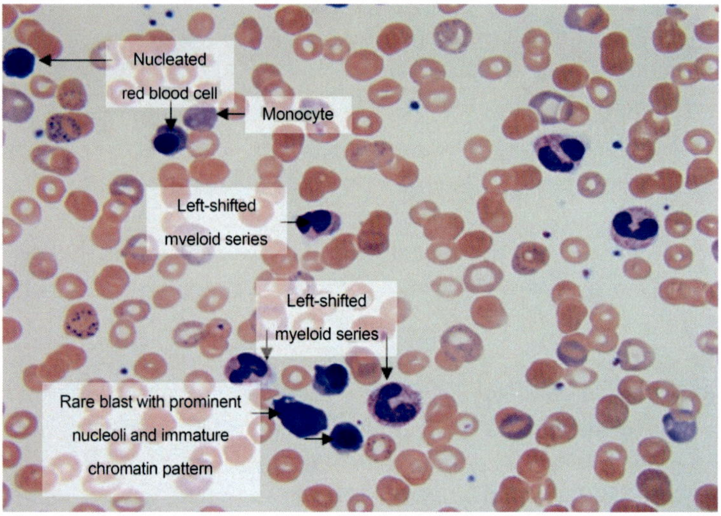

Figure 1.17. Sample blood smears of hematological abnormalities from COVID-19 patients requiring intensive care unit (ICU) admission

Source: Lin E, Araj E, Markantonis J, Luu H, & Chen M., (2022). hematologic complications in patients hospitalized with COVID-19 infection. Hematology Reports. 2022; 14(3), 228–234. https://doi.org/10.3390/hematolrep14030031: © 2022 by the authors. Licensee MDPI, Basel, Switzerland. This article is an open access article distributed under the terms and conditions of the Creative Commons Attribution (CC BY) license (https://creativecommons.org/licenses/by/4.0/).

1.8.3. Neutrophils

Studies show that neutrophilia corresponds to a hyperinflammatory state and elevated levels of cytokines are the pathogenic mechanism in COVID-19 patients. Neutrophils are associated with several viral respiratory conditions related to ARDS (acute respiratory distress syndrome) (Camp & Jonsson, 2017). In severe cases, some patients showed neutrophilia supported by leukocytosis (Yamada et al., 2020). It was observed that the number of circulating neutrophils increased gradually as COVID-19 progressed in patients, making neutrophilia a marker for critical respiratory diseases (Zhang et al., 2020). The markers, neutrophils, and leukocytes were remarkably higher in critical patients, and their levels elevated as the condition further progressed. Studies showed that thrombocyte count was the key to the confirmation and prognosis of diseases. Although the levels of lymphocyte, CRP, and D-dimer were not indicative of the diagnostic value, they were indicative of the intensity of COVID-19 (Sorya&Ulhaq, 2020).

Morphological modifications in the circulating neutrophils like the dense presence of granules in the cytoplasm and reduction of nuclear lobularity were observed in the peripheral blood smears. These alterations were found to be temporary and reversible prior to the emergence of large atypical reactive lymphocytes which are the characteristic trait of viral infections (Nazarullah et al., 2020; Zing et al., 2020).

Elevated levels of neutrophils were observed in the lung tissues along with their increased levels in the blood. Elevated diffusion of dysfunctional and/or immature neutrophils signifies an abnormal immune response in the critical patients. As seen in other viral infections, the COVID-19 virus also stimulates the release of extracellular neutrophils which advances to tissue damage. Noticeable activation of neutrophils can worsen the patient's response to the disease. Furthermore, autopsies of the lungs showed lung capillaries and alveolar spaces to be populated with neutrophils. Through increased cytokine production and degranulation, neutrophils play an essential role as stimulators for hyperinflammation in COVID-19 patients. Median neutrophil count measured in severe cases was higher compared to the moderate or milder groups of patients affected with COVID-19 and a noticeable elevated level in neutrophil count was observed from day 7 to 9 post the onset of the symptom. Furthermore, deciphering the association between poor outcome and neutrophilia, alterations in the neutrophil count in tissues or peripheral blood samples were found to be related to damage of the lungs in the Covid-19 patients (Palladino, 2021).

1.8.4. Ratio of Neutrophil-Lymphocyte and Platelet-Lymphocyte

The NLR (absolute neutrophil count/absolute lymphocyte count ratio) and PLR (absolute platelet count/absolute lymphocyte count ratio) prove to be essential in the diagnosis, follow-ups, and examination of several systemic inflammatory responses like cholangiocarcinoma, acute pancreatitis, ischemic heart disease, along with being a prognostic marker for malignant tumors (Palladino, 2021). The levels of PLR and NLR correspond with the severity of COVID-19 in patients. Patients at critical levels showed higher values of NLR and PLR compared to the mild and moderate groups. The NLR is found to be increased in the bloodstream of COVID-19-positive patients. Studies also proved that compared to the neutrophil count alone, NLR clubbed together with IgG is a better indicator for determining the severity of COVID-19 in patients (Zhang et al., 2020). The values of PLR and NLR are separate factors corresponding with the progression of COVID-19, and their exact mechanism is not understood. The initial stage of the disease shows the overall leukocyte count in peripheral blood to be normal or even a decline and the lymphocyte count to decline. Studies in deceased patients infected with COVID-19 showed an elevated level in the initial and peak NLR values compared to the survivors ($P < 0.001$). Elevation in NLR indicates the continuous escalation in neutrophils associated with a reduction in lymphocytes. An increase in NLR signifies the presence of a bacterial infection whereas the decline in lymphocytes indicates a compromised health condition, thus suggesting the priority given to patients infected with COVID-19 with elevated NLR. Detrimental outcomes were observed in patients with an elevation in platelet count in the course of the disease. Also, the PLR during that time was considered as a prognostic factor for prolonged hospitalization. Studies by Lu and Wang describe a case report where fluctuations in the NLR and PLR values were observed during the entire progressive phase of Covid-19. Levels of WBC, platelets, neutrophils, and NLR were observed to increase gradually from the day of admission through their peak on day 14. On day 9, the PLR was observed to have attained its peak and the lymphocyte count was not at its peak but showed a rise. On day 14, where the patient's health had improved, the PLR and NLR had returned to normal (Lu & Wang, 2020). Studies by Bustug revealed elevated levels of ferritin and PLR, both independent factors describing the severity of the disease (Bustug et al., 2020). Studies also showed PLR to be associated with inflammation and this could be indicative of mortality in patients on hemodialysis. Covid-19 virus, SARS-COV-2 stimulated hyperinflammation was elevated PLR suggestive of severe prognosis (Ahbap et al., 2016).

NLR and PLR are easily obtained from serum CBC with a differential profile and indicate the presence of relative neutrophilia, lymphopenia, and thrombocytosis. Inflammatory markers such as ESR, CRP, ferritin, lactate dehydrogenase, and PCT are regularly checked in Covid-19 patients. PLR and NLR can also be easily calculated through a differential count, making this a cost-effective technique in many third-world countries. Observing these predictors that indicate the severity of the disease may help clinicians keep track and follow up with patients who are at higher risk for disease progression (Palladino, 2021).

1.8.5. Monocytes

Monocytes remain in circulation for about 1–2 days after which they differentiate into macrophages resident in specific tissues, comprising about 5–9% of the total peripheral leukocyte count. Through the action of ACE-2 (angiotensin-converting enzyme), the COVID virus infects the CD14+ monocytes; however, the replication of viruses in them is undetectable or generally low (Zhang et al., 2021). Monocytes infected with the COVID-19 virus are known to produce large numbers of inflammatory mediators that support the SARS-Cov-2-associated cytokine storm. Studies have discovered that patients in their early recovery stage had a principal subset of CD14++ IL1β+ monocytes. Studies by Zhang and team found no remarkable difference in the absolute count of monocytes in COVID-19-infected patients and healthy individuals. However, they did notice elevated levels of activated macrophages/monocytes in COVID-19 patients along with noticeable functional and morphological differences that are indicative of patients needing ICU and longer hospitalization. It is observed that the number of CD14++CD16- (circulating monocytes) declines and the number of CD14++CD16+ (intermediate monocytes) and CD14++CD16++ (non-classical monocytes) elevate in COVID-19 positive patients. Furthermore, patients with this disease exhibit larger than normal monocytes that are easily detected on FSC (forward scatter) analysis performed on routine flow cytometry along with monocytes with high FSC. The high FSC monocytes are CD14+, CD11b, CD68+, CD206+, CD16+, CD80+, and CD163+ and they release IL-10, IL-6, and TNF-alpha in accordance with the inflammatory response. Patients with elevated levels of normal monocytes exhibit a better prognosis combined with speedy recovery and discharge from the hospital.

These discoveries are attributed to COVID-19 specifically as they are not detected in patients infected with other viral infections like HIV (human immunodeficiency virus) and H1N1 influenza (Zhang et al., 2021). A remarkable increase in CD14+CD16+ monocytes that produce IL-6 was noticed in the peripheral blood samples of severe and ICU COVID-19-positive patients compared to those with mild symptoms. These monocytes from positive patients secreted GM-CSF (granulocyte-macrophage colony-stimulating factor). GM-CSF-responsive monocytic cells need GM-CSF to stimulate damage in tissues, hence it is linked to be responsible for the inflammatory response of cytokine storms in COVID-19 infection (Zhou et al., 2020). SARS-CoV-2 infection in monocytes hinders the anti-viral adaptive immune responses. Hence, the infection not only interferes with the monocytes but also causes the stimulation of cytokine production and cytokine-mediated signaling pathways, resulting in reduced systemic hyperinflammation (Palladino, 2021).

1.8.6. Basophils

Basophils are known to exit circulation and travel to sites of inflammation in an allergic infection and inflammation while attaching antigens on their surfaces, thus strengthening the immunological memory responses. Secretion of IgA or IgG by B cells is amplified by stimulated basophils.

Studies showed a severe decline in the basophils in acute and severe stages of COVID-19, indicating the degree of depletion of basophils to be affected by the efficacy of IgG response to the virus (Rodriguez et al., 2020). Compared to healthy controls, the basophil count was low in patients suffering from influenza and COVID-19. This lower count was noticed in the early stages of the disease. Eosinophils and basophils make IL-4, an essential cytokine, which triggers the proliferation of stimulated B and T cells. Hence, a decline in the eosinophil and basophil count in COVID-19 patients can aid in further explaining the decline in the lymphocyte count. Studies on 452 critical COVID-19 patients by Qin et al., found higher neutrophil and leukocyte counts and lower basophil counts (Qin et al., 2020). Low basophil counts are linked with viral infections in patients who have a compromised immune system. Studies have shown that various HIV proteins can correspond with different receptors on the human basophils and HIV-infected basophils were observed in the peripheral blood samples from patients with immunodeficiency syndrome (Palladino, 2021). Considering these factors, a technique in which the virus encapsulates in the basophils can be assumed and could be a leading study for the low basophil count in COVID-19-positive patients.

1.8.7. Eosinophils

The contribution and role of eosinophils in COVID-19 are not known. Physiologically, the gastrointestinal tract and lungs have tissue-resident eosinophils that provide immunity, metabolism, and organ growth. Studies by Zhang and team showed that 53% of 140 COVID-19-positive patients in Wuhan had eosinopenia when admitted (Zhang et al., 2020). Studies by Mu and team further reported noticeable eosinopenia in 72 out of 95 positive patients, where the eosinophil percentage was 0.3% and the absolute eosinophil count was 0.01*10^9/L. It was also observed that the eosinophil blood count gradually came back to normal when the patients healed, and it continued to reduce in unhealed patients (Mu et al., 2021). Similar conclusions were inferred by Tan and team in a study of 33 out of 40 patients, where the eosinophils reduced with the severity of the disease and returned to normal upon healing, as shown by Spearman's correlation coefficient (rs = -0.462; P = 0.003), making eosinophils a factor for predicting the diagnosis of the patient (Tan et al., 2021).

Zappert first mentioned the condition of eosinopenia, which is a response to infection(s), and its pathophysiology in COVID-19 needs to be studied in detail. However, it can be stipulated to be multifactorial and correlates to the movement of eosinophils to the site of inflammation. Also, it is associated with the inhibition of eosinophil circulation from the bone marrow, sieging of eosinophilopoiesis, direct eosinophil apoptosis that is stimulated by IFN-1 released during the acute phase of inflammation, and decline in the expression of chemokine receptors that are the adhesion molecules (Palladino, 2021).

In order to show an example, the findings on the alterations in blood cell counts from a multicenter cross-sectional analysis by Chowdhury et al (2021) have been shared Table 1.10 as documented by the researchers.

Table 1.10. The alterations in blood cell counts in COVID-19 from a multicenter cross-sectional analysis by Chowdhury et al (2021)

Source: Mohiuddin Chowdhury ATM, Karim MR, Ali MA, Islam J, Li Y & He S (2021) Clinical Characteristics and the Long-Term Post-recovery Manifestations of the COVID-19 Patients—A Prospective Multicenter Cross-Sectional Study. Front. Med., 8, 663670. doi: 10.3389/fmed.2021.663670: Copyright © 2021 Mohiuddin Chowdhury, Karim, Ali, Islam, Li and He. This is an open-access article distributed under the terms of the Creative Commons Attribution License (CC BY). The use, distribution or reproduction in other forums is permitted, provided the original author(s) and the copyright owner(s) are credited and that the original publication in this journal is cited, in accordance with accepted academic practice. No use, distribution or reproduction is permitted which does not comply with these terms.

1.8.8. RBCs

A remarkable influence of SARS-CoV-2 infection on the structural membrane homeostasis of RBCs at the lipid and protein levels was studied. RBCs from COVID-19 infected patients showed elevated levels of glycolytic intermediates coupled with fragmentation and oxidation of membrane proteins. This signifies the impact of Covid-19 on two important mechanisms that regulate the membranes of RBCs and the affinity of Hb towards oxygen. The RBCs from COVID-19 patients may not be sufficient to correspond to the surrounding diversity in Hb-oxygen saturation during its movement from the lungs to the bloodstream, and it can also have a compromised capacity to carry and deliver oxygen (Thomas et al., 2020). The levels of NO (nitric oxide) are high in the RBCs of COVID-19 patients compared to non-Covid-19 hypoxemic patients. However, the process(es) resulting from the buildup of intracellular NO in the RBCs of these patients is not yet clear. Also, AIHA (autoimmune hemolytic anemia) was studied in these patients. AIHA results in the death of platelets, and RBCs can alter the action of platelets via RBC-platelet interactions or chemical signaling. This could prove hemolysis to account for microvascular coagulation observed in the patients. Studies by Berzuini and team outlined that nearly half of COVID-19 patients at their center displayed a positive DAT (direct antiglobulin test). Also, the sample did not react with any other cells except the RBCs from COVID-19 patients that were exhibited as DAT negative. This indicates that COVID-19 can alter the RBC membrane and this can prove classic and new antigenic epitopes (Palladino, 2021).

Table 1.11. The major alterations in hematological markers in COVID-19 exhaustively summarized by Lechuga et al, (2023)

Hematological parameter	Status in long COVID
Cells and cellular parameters	
RBC	↓(Kubánková et al., 2021; Grau et al., 2022; Lin et al., 2022)
MCV	↑(Lin et al., 2022)
MCHC	↑(Alfadda et al., 2022)
Lymphocytes	↓ (Mandal et al., 2021; Moreno-Pérez et al., 2021; Alfadda et al., 2022); ↔(Darcis et al., 2021)
Blood components and proteins	
Hemoglobin	↓(Kubánková et al., 2021; Pasini et al., 2021; Grau et al., 2022; Pereira-Roche et al., 2022); ↔ (Darcis et al., 2021; Alfadda et al., 2022)
Platelet count	↔ (Mandal et al., 2021; Pasini et al., 2021; Alfadda et al., 2022)
D-dimer	↑ (Mandal et al., 2021; Moreno-Pérez et al., 2021; Pasini et al., 2021; Fan et al., 2022; Kalaivani and Dinakar, 2022)
Ferritin	↑ (Moreno-Pérez et al., 2021; Pasini et al., 2021); ↔(Darcis et al., 2021; Mandal et al., 2021; Alfadda et al., 2022; Pérez-González et al., 2022)
C-reactive protein (C.R.P.)	↑ (Mandal et al., 2021; Pasini et al., 2021; Pérez-González et al., 2022); ↔ (Darcis et al., 2021; Kalaivani and Dinakar, 2022)
Lactate dehydrogenase	↔(Moreno-Pérez et al., 2021; Pasini et al., 2021; Alfadda et al., 2022)
Cytokines	
IL-6	↑ (Fan et al., 2022; Schultheiß et al., 2022); ↓(Williams et al., 2022); ↔ (Kalaivani and Dinakar, 2022; Queiroz et al., 2022)
TNF-α	↑(Peluso et al., 2021; Schultheiß et al., 2022); ↔ (Queiroz et al., 2022; Williams et al., 2022)
IL-1β	↑(Schultheiß et al., 2022); ↔ (Williams et al., 2022)
IL-2	↓ (Williams et al., 2022); ↑ (Queiroz et al., 2022)
IL17	↓ (Williams et al., 2022); ↑(Queiroz et al., 2022)
IFNγ	↓(Williams et al., 2022); ↔ (Queiroz et al., 2022)
IFN-β	↑(Phetsouphanh et al., 2022)
IFN-λ1	↑(Phetsouphanh et al., 2022)
IL-8	↓(Williams et al., 2022); ↔ (Schultheiß et al., 2022); ↑(Phetsouphanh et al., 2022)
IL-10	↔ (Williams et al., 2022); ↓(Queiroz et al., 2022)
IL-4	↓(Queiroz et al., 2022; Williams et al., 2022); ↔(Schultheiß et al., 2022)

Source: Lechuga GC, Morel CM, & De-Simone SG, (2023) Hematological alterations associated with long COVID-19. Front. Physiol., 14, 1203472. doi: 10.3389/fphys.2023.1203472: Copyright © 2023 Lechuga, Morel and De-Simone. This is

Samples from COVID-19 patients revealed a lower Hb concentration and pathological elevation in the ferritin concentrations. Studies by Wang and team reported a < 110 g/L decrease in Hb concentration in 19% (age range 33–53) and Huang and team showed this decrease to be 38% (age range 26–88) of hospitalized patients (Wang et al., 2020; Huang et al., 2020). This difference in lower Hb percentage can be attributed to the age difference in both the studies and the Hb concentration lower than 110 g/L was associated with the progression of the disease. Another study examined data from a sample size of 189 studies with 57,563 positive patients spread across all age groups where the mean Hb levels were 129.7 g/L and this value was further reduced with older age and the severity of COVID-19. Most patients, especially elderly, males, and patients with hypertension showed higher concentrations of ferritin (Taneri et al., 2020).

Studies based on 24,262 COVID-19-positive patients showed a mean ferritin concentration of 777.33 ng/mL, and this mean was higher corresponding to the severity of the disease and the older age group. The mean concentrations of serum ferritin were found to be higher in non-survivors (562 ng/mL) compared to the survivor group (493 ng/mL), and these concentrations were observed to increase along with the severity of the disease (Zhou et al., 2020). Hyperferritinemia and anemia were found to be closely associated with mortality despite any pathological conditions. The Hb values were lower while the ferritin concentrations were higher in old age among those with hypertension and diabetes and those admitted to the ICU. The major alterations in hematological markers in COVID-19 were exhaustively summarized by Lechuga et al. (2023) and are shown in Table 1.11.

Anemia associated with COVID-19 was attributed to the pattern of sideroblastic anemia that emerges from modifications in iron metabolism, and elevated ferritin levels could be the indication for the underlying inflammatory response to the viral attack found in the patients. Increased ferritin concentrations were found to correlate with prolonged hospital stay followed by an increased risk for the need for mechanical ventilation and admission to the ICU. Inflammation in COVID-19 is marked by the biomarkers of iron metabolism and Hb. Anemic patients showed noticeably elevated levels of IL-6, CRP, and PCT, where Hb was linked to IL-6 and PCT concentrations. Patients with iron deficiency showed extremely high concentrations of PCT, IL-6, and CRP compared to patients with no iron deficiency, and ferritin was found to be positively associated with these levels. These higher concentrations are in conformity with the correlation of the altered iron homeostasis linked with severe inflammation resulting in severe injury to the lungs and respiratory failure. The inflammation caused by altered iron homeostasis and anemia is an essential clinical marker to determine the severity of infection in COVID-19 patients.

An extreme level of acute respiratory failure in coronavirus 2 was observed to interfere with Hb at the bone marrow and erythrocyte level; it is known that the new variant of SARS-CoV-2 attacks the bone marrow erythroblasts (Majeed &Shajar, 2020). Studies also revealed elevated RDW in patients indicating the presence of erythroid myelodysplasia as these levels are higher when the cells produced are immature (Foy et al., 2020). The Covid-19 virus interacts via CD26, ACE2, and CD147 among other receptors present on the precursors of blood cells and erythrocytes resulting in viral endocytosis. This ultimately results in COVID-19 viruses attacking the heme on the 1-beta chain of Hb stimulating hemolysis, creating a complex coupled with the released heme and producing dysfunctional Hb (Wenzhong&Hualan, 2021).

1.9. A FINAL WORD

The pandemic disease Coronavirus of 2019 has noteworthy implications in the hematopoietic system. Common hematological abnormalities are identified in COVID-19 patients. From the initial stage, lymphocytes, platelets, hemoglobin, basophils, and eosinophils showed a decrease in their count, which was directly correlated with the severity of the disease. The impact on the kinetics of monocytes in the disease is not yet clear. This could be due to SARS-CoV-2 infection in monocytes weakening the antiviral immune response. An increase in neutrophils and markers like PLR and NLR also correlate with the stages of disease progression. Careful examination of laboratory test evidence at the baseline and during different progressive levels of the disease may help clinicians devise a tailored treatment that can promptly provide intensive care to patients in an emergency. Further research is needed to describe the relationship between hematological responses and disease progression, as well as the outcomes of the disease, in order to provide accurate and timely assistance to patients, treating them in the early stages and preventing them from reaching the life-threatening stage of the disease.

"Gratitude works its magic by serving as an antidote to negative emotions. It's like white blood cells for the soul, protecting us from cynicism, entitlement, anger, and resignation."

— Arianna Huffington

REFERENCES

1. Ahbap, E., Sakaci, T., Kara, E., Sahutoglu, T., Koc, Y., Basturk, T., &Kayalar, A. O. (2016). Neutrophil-to-lymphocyte ratio and platelet-to-lymphocyte ratio in evaluation of inflammation in end-stage renal disease. *Clinical Nephrology, 85*, 199–208.
2. Ahmed, I., Majeed, A., & Powell, R. (2007). Heparin induced thrombocytopenia: Diagnosis and management update. *Postgraduate Medical Journal, 83*(983), 575–582.
3. Al-Saadi, E. A. K. D., &Abdulnabi, M. A. (2022). Hematological changes associated with COVID-19 infection. *Journal of Clinical Laboratory Analysis, 36*(1), e24064. https://doi.org/10.1002/jcla.24064.
4. Arnold, D. M., Patriquin, C. J., &Nazy, I. (2017). Thrombotic microangiopathies: A general approach to diagnosis and management. *Canadian Medical Association Journal, 189*(4), E153–E159. https://doi.org/10.1503/cmaj.160142.
5. Barbalato, L., &Pillarisetty, L. S. (2022). Histology, Red Blood Cell. In *StatPearls*. StatPearls Publishing.
6. Barcia, J. J. (2007). The Giemsa stain: Its history and applications. *International Journal of Surgical Pathology, 15*(3), 292–296.
7. Bastug, A., Bodur, H., Erdogan, S., Gokcinar, D., Kazancioglu, S., Kosovali, B. D., &Duman, H. (2020). Clinical and laboratory features of COVID-19: Predictors of severe prognosis. *International Immunopharmacology, 88*, 106950. https://doi.org/10.1016/j.intimp.2020.106950.
8. Biswajit, H., Pratim, P. P., Kumar, S. T., Shilpi, S., Krishna, G. B., & Aditi, A. (2012). Aplastic anemia: A common hematological abnormality among peripheral pancytopenia. *North American Journal of Medical Sciences, 4*(9), 384–388.
9. Bleeker, J. S., &Hogan, W. J. (2011). Thrombocytosis: Diagnostic evaluation, thrombotic risk stratification, and risk-based management strategies. *Thrombosis, 2011*, 536062. https://doi.org/10.1155/2011/536062.
10. Camp, J. V., & Jonsson, C. B. (2017). A role for neutrophils in viral respiratory disease. *Frontiers in Immunology, 8*, 550. https://doi.org/10.3389/fimmu.2017.00550.
11. Cossarizza, A., De Biasi, S., Guaraldi, G., Girardis, M., & Mussini, C. for the Modena Covid-19 Working Group (MoCo19). (2020). SARS-CoV-2, the virus that causes COVID-19: Cytometry and the new challenge for global health. *Cytometry Part A, 97*, 340–343. https://doi.org/10.1002/cyto.a.24002.

12. Das, S., Abreu, C., Harris, M., Shrader, J., &Sarvepalli, S. (2022). Severe thrombocytopenia associated with dengue fever: An evidence-based approach to management of thrombocytopenia. *Case Reports in Hematology, 2022*, 3358325. https://doi.org/10.1155/2022/3358325.

13. David, A. N., Jinadu, M. Y., Wapmuk, A. E., Gbajabiamila, T. A., Okwuzu, J. O., Herbertson, E. C., &Ezechi, O. C. (2018). Prevalence and impact of sickle cell trait on the clinical and laboratory parameters of HIV infected children in Lagos, Nigeria. *Pan African Medical Journal, 31*, 113.

14. Dotson, J. L., &Lebowicz, Y. (2022). Myelodysplastic Syndrome. In *StatPearls*. StatPearls Publishing.

15. Foy, B. H., Carlson, J. T., Reinertsen, E., Padros Valls, R., Pallares Lopez, R., Palanques-Tost, E., et al. (2020). Elevated RDW is associated with increased mortality risk in COVID-19. *JAMA Network Open, 3*, e2022058.

16. Galanello, R., &Origa, R. (2010). Beta-thalassemia. *Orphanet Journal of Rare Diseases, 5*, 11. https://doi.org/10.1186/1750-1172-5-11.

17. Giannakeas, V., Kotsopoulos, J., Brooks, J. D., Cheung, M. C., Rosella, L., Lipscombe, L., &Narod, S. A. (2022). Platelet count and survival after cancer. *Cancers (Basel), 14*(3), 549. https://doi.org/10.3390/cancers14030549.

18. Häggström, M. (2014). Medical gallery of Mikael Häggström 2014. *WikiJournal of Medicine, 1*(2). https://en.wikiversity.org/wiki/WikiJournal_of_Medicine/Medical_gallery_of_Mikael_H%C3%A4ggstr%C3%B6m_2014 (accessed on 22 June 2024).

19. Hill, A., & Hill, Q. A. (2018). Autoimmune hemolytic anemia. *Hematology American Society of Hematology Education Program, 2018*(1), 382–389. https://doi.org/10.1182/asheducation-2018.1.382.

20. Huang, Y., Tu, M., Wang, S., Chen, S., Zhou, W., Chen, D., et al. (2020). Clinical characteristics of laboratory confirmed positive cases of SARS-CoV-2 infection in Wuhan, China: A retrospective single center analysis. *Travel Medicine and Infectious Disease, 36*, 101606. https://doi.org/10.1016/j.tmaid.2020.101606.

21. K T N, Prasad, K., & Singh, B. M. K. (2022). Analysis of red blood cells from peripheral blood smear images for anemia detection: A methodological review. *Medical & Biological Engineering & Computing, 60*(9), 2445–2462.

22. Kabat, G. C., Kim, M. Y., Manson, J. E., Lessin, L., Lin, J., Wassertheil-Smoller, S., & Rohan, T. E. (2017). White blood cell counts and total and cause-specific mortality in the Women's Health Initiative. *American Journal of Epidemiology, 186*(1), 63–72.

23. Karpman, D., Manea, M., Vaziri-Sani, F., Ståhl, A. L., &Kristoffersson, A. C. (2006). Platelet activation in hemolytic uremic syndrome. *Seminars in Thrombosis and Hemostasis, 32*(2), 128–145.

24. Köhne, E. (2011). Hemoglobinopathies: Clinical manifestations, diagnosis, and treatment. *DeutschesÄrzteblatt International, 108*(31–32), 532–540. https://doi.org/10.3238/arztebl.2011.0532.

25. Lippi, G., Plebani, M., & Henry, B. M. (2020). Thrombocytopenia is associated with severe coronavirus disease 2019 (COVID-19) infections: A meta-analysis. *ClinicaChimica Acta, 506*, 145–148. https://doi.org/10.1016/j.cca.2020.03.022.

26. Lu, G., & Wang, J. (2020). Dynamic changes in routine blood parameters of a severe COVID-19 case. *Clinical Chemistry and Laboratory Medicine (CCLM), 58*, 98–102.

27. Lu, X., & Chang, R. (2023). Polycythemia Vera. In *StatPearls*. StatPearls Publishing.

28. Luu, S., Gardiner, E. E., & Andrews, R. K. (2018). Bone marrow defects and platelet function: A focus on MDS and CLL. *Cancers (Basel), 10*(5), 147. https://doi.org/10.3390/cancers10050147.

29. Majeed, A., &Shajar, M. A. (2020). Is hemoglobin the missing link in the pathogenesis of COVID-19? *Anaesthesia, Pain & Intensive Care, 24*, 9–12.

30. Mohammad, A. A. (2018). Myelodysplastic syndrome from theoretical review to clinical application view. *Oncology Reviews, 12*(2), 397.

31. Moore, C. A., & Krishnan, K. (2023). Aplastic Anemia. In *StatPearls*. StatPearls Publishing.

32. Mu, T., Yi, Z., Wang, M., Wang, J., Zhang, C., Chen, H., & Wang, D. (2021). Expression of eosinophil in peripheral blood of patients with COVID-19 and its clinical significance. *Journal of Clinical Laboratory Analysis, 35*, e23620. https://doi.org/10.1002/jcla.23620.

33. Nah, E. H., Kim, S., Cho, S., & Cho, H. I. (2018). Complete blood count reference intervals and patterns of changes across pediatric, adult, and geriatric ages in Korea. *Annals of Laboratory Medicine, 38*(6), 503–511. https://doi.org/10.3343/alm.2018.38.6.503.

34. Nazarullah, A., Liang, C., Villarreal, A., Higgins, R. A., &Mais, D. D. (2020). Peripheral blood examination findings in SARS-CoV-2 infection. *American Journal of Clinical Pathology, 154*, 319–329.

35. O'Farrill-Santoscoy, F., O'Farrill-Cadena, M., & Fragoso-Morales, L. E. (2013, July). Evaluación del tratamiento a mujeresembarazadas con

anemia ferropénica [Evaluation of treatment of iron deficiency anemia in pregnancy]. *Ginecología y Obstetricia de México, 81*(7), 377–381. [Article in Spanish].

36. Palladino, M. (2021). Complete blood count alterations in COVID-19 patients: A narrative review. *Biochemia Medica (Zagreb), 31*(3), 030501. https://doi.org/10.11613/BM.2021.030501.

37. Pillai, A. A., Fazal, S., Mukkamalla, S. K. R., &Babiker, H. M. (2023). Polycythemia. In *StatPearls*. StatPearls Publishing.

38. Prinyakupt, J., &Pluempitiwiriyawej, C. (2015). Segmentation of white blood cells and comparison of cell morphology by linear and naïve Bayes classifiers. *BioMedical Engineering OnLine, 14*, 63.

39. Qin, C., Zhou, L., Hu, Z., Zhang, S., Yang, S., Tao, Y., & Jiang, S. (2020). Dysregulation of immune response in patients with COVID-19 in Wuhan, China. *Clinical Infectious Diseases, 71*, 762–768. https://doi.org/10.1093/cid/ciaa248.

40. Rodriguez, L., Pekkarinen, P. T., Lakshmikanth, T., Tan, Z., Consiglio, C. R., Pou, C., & Bjorkman, L. (2020). Systems-level immunomonitoring from acute to recovery phase of severe COVID-19. *Cell Reports Medicine, 1*, 100078. https://doi.org/10.1016/j.xcrm.2020.100078.

41. Schwartz, R. S., & Conley, C. L. (2023). Blood. *Encyclopedia Britannica.* https://www.britannica.com/science/blood-biochemistry (accessed on 22 June 2024).

42. Sedrak, A., &Kondamudi, N. P. (2023). Sickle Cell Disease. In *StatPearls*. StatPearls Publishing.

43. Seyoum, M., Enawgaw, B., &Melku, M. (2018). Human blood platelets and viruses: Defense mechanism and role in the removal of viral pathogens. *Thrombosis Journal, 16*, 16.

44. Shastry, I., &Belurkar, S. (2019). The spectrum of red blood cell parameters in chronic kidney disease: A study of 300 cases. *Journal of Applied Hematology, 10*, 61–66.

45. Soraya, G. V., &Ulhaq, Z. S. (2020). Crucial laboratory parameters in COVID-19 diagnosis and prognosis: An updated meta-analysis. *MedicinaClínica (English Edition), 155*, 143–151.

46. Tan, Y., Zhou, J., Zhou, Q., Hu, L., & Long, Y. (2021). Role of eosinophils in the diagnosis and prognostic evaluation of COVID-19. *Journal of Medical Virology, 93*, 1105–1110.

47. Taneri, P. E., Gómez-Ochoa, S. A., Llanaj, E., Raguindin, P. F., Rojas, L. Z., Roa-Díaz, Z. M., et al. (2020). Anemia and iron metabolism in

COVID-19: A systematic review and meta-analysis. *European Journal of Epidemiology, 35*, 763–773. https://doi.org/10.1007/s10654-020-00678-5.

48. Thomas, T., Stefanoni, D., Dzieciatkowska, M., Issaian, A., Nemkov, T., Hill, R. C., & Hansen, K. C. (2020). Evidence of structural protein damage and membrane lipid remodeling in red blood cells from COVID-19 patients. *Journal of Proteome Research, 19*, 4455–4469.
49. Tigner, A., Ibrahim, S. A., & Murray, I. V. (2022). Histology, White Blood Cell. In *StatPearls*. StatPearls Publishing.
50. Triscott, J. A., Dobbs, B. M., McKay, R. M., Babenko, O., &Triscott, E. (2015). Prevalence and types of anemia and associations with functional decline in geriatric inpatients. *Journal of Frailty and Aging, 4*(1), 7–12.
51. Turner, J., Parsi, M., &Badireddy, M. (2023). Anemia. In *StatPearls*. StatPearls Publishing.
52. Wang, L., Duan, Y., Zhang, W., Liang, J., Xu, J., Zhang, Y., & Wang, M. (2020). Epidemiologic and clinical characteristics of 26 cases of COVID-19 arising from patient-to-patient transmission in Liaocheng, China. *Clinical Epidemiology, 12*, 387–391.
53. Wenzhong, L., &Hualan, L. (2021). COVID-19: Captures iron and generates reactive oxygen species to damage the human immune system. *Autoimmunity, 54*, 213–224.
54. Wu, Z., & McGoogan, J. M. (2020). Characteristics of and important lessons from the coronavirus disease 2019 (COVID-19) outbreak in China: Summary of a report of 72,314 cases from the Chinese Center for Disease Control and Prevention. *JAMA, 323*, 1239–1242. https://doi.org/10.1001/jama.2020.2648.
55. Yamada, T., Wakabayashi, M., Yamaji, T., Chopra, N., Mikami, T., Miyashita, H., & Murakami, K. (2020). Value of leukocytosis and elevated C-reactive protein in predicting severe coronavirus 2019 (COVID-19): A systematic review and meta-analysis. *Clinical Chemistry and Laboratory Medicine (CCLM), 58*, 235–243.
56. Zhang, B., Zhou, X., Zhu, C., Song, Y., Feng, F., Qiu, Y., & Zhang, H. (2020). Immune phenotyping based on the neutrophil-to-lymphocyte ratio and IgG level predicts disease severity and outcome for patients with COVID-19. *Frontiers in Molecular Biosciences, 7*, 157. https://doi.org/10.3389/fmolb.2020.00157.
57. Zhang, D., Guo, R., Lei, L., Liu, H., Wang, Y., Wang, Y., & Cao, X. (2021). COVID-19 infection induces readily detectable morphological and

inflammation-related phenotypic changes in peripheral blood monocytes. *Journal of Leukocyte Biology, 109*, 13–22.

58. Zhang, J. J., Dong, X., Cao, Y. Y., Yuan, Y. D., Yang, Y. B., Yan, Y. Q., &Xie, J. (2020). Clinical characteristics of 140 patients infected with SARS-CoV-2 in Wuhan, China. *Allergy, 75*, 1730–1741. https://doi.org/10.1111/all.14238.
59. Zhang, L., Huang, B., Xia, H., Fan, H., Zhu, M., Zhu, L., & Wang, W. (2020). Retrospective analysis of clinical features in 134 coronavirus disease 2019 cases. *Epidemiology and Infection, 148*, e199.
60. Zhao, Q., Liu, R., Chen, H., Yang, X., Dong, J., & Bai, M. (2023). White blood cells and coronary heart disease: A mendelian randomization study. *Frontiers in Genetics, 14*, 1127820. https://doi.org/10.3389/fgene.2023.1127820.
61. Zhou, F., Yu, T., Du, R., Fan, G., Liu, Y., Liu, Z., et al. (2020). Clinical course and risk factors for mortality of adult inpatients with COVID-19 in Wuhan, China: A retrospective cohort study. *The Lancet, 395*, 1054–1062. https://doi.org/10.1016/S0140-6736(20)30566-3.
62. Zhou, Y., Fu, B., Zheng, X., Wang, D., Zhao, C., Qi, Y., & Wang, F. (2020). Pathogenic T-cells and inflammatory monocytes incite inflammatory storms in severe COVID-19 patients. *National Science Review, 7*, 998–1002. https://doi.org/10.1093/nsr/nwaa041.
63. Zini, G., Bellesi, S., Ramundo, F., & D'Onofrio, G. (2020). Morphological anomalies of circulating blood cells in COVID-19. *American Journal of Hematology, 95*, 870–872.

Chapter 2

BLOOD CELLS AND THEIR INTERACTIONS

"Hemoglobin is complicated enough to be interesting. I began research in the nature of hemoglobin in 1935, and now 31 years later, I am still entranced by this wonderful molecule."

—Linus Pauling. "Science and World Problems." Speech sponsored by the Carl Neuberg Society for International Scientific Relations, New York City, NY March 30, 1966.

2.1. INTRODUCTION

The unique fluid dynamics of blood have been uncovered through a long history of intensive research (Munn &Dupin, 2008). Blood's particulate nature causes its hematocrit and viscosity to vary within the microvasculature, as pioneers discovered in the early 20th century (Cokelet& Goldsmith, 1991). As a result of RBC deformation and reversible aggregate formation, others have shown that the viscosity of the vasculature is shear- and conduit-dependent, thus minimizing flow resistance (Bishop et al., 2001). Studies have revealed details about how individual blood cells flow and interact mechanically with fluids and one another (Simon & Goldsmith, 2002). In addition, Simon and Goldsmith (2002) examined the interactions between immune cells and vessel walls, or the dynamics of platelet aggregation (Kroll et al., 1996). As a result of all this work, blood has been demonstrated to be an organized, yet dynamic, collection of cells interacting with each other and with their conduits. It is important to consider blood as an organ in its own right. Physiology (i.e., how the blood flow interacts with other organs (for example, by interacting with the shear force exerted on the vessel wall) is crucial for understanding bulk blood flow because it depends on multiple factors, including cell biology, cell mechanics, "tissue" biophysics, and physiology.

New technologies have historically driven advances in blood flow dynamics. There have been several important advances made in intravital microscopy (Brown et al., 2001), video cinematography, and mathematical modeling (Das et al., 2000). Although these approaches have provided insights into the rheology of individual cells and bulk flow properties of blood, much remains to be understood about blood at the mesoscale, i.e., how cells interact with tissues.

Prior to the 19th century, hemorrhagic syndrome was one of the leading causes of death in prehistoric times. In the second half of the 19th century, coagulation was discovered and fibrin, previously called plasmin, was identified. Until the beginning of the twentieth century, platelets were largely unknown as a factor in bleeding arrest. Apart from wars, thrombosis has been a leading cause of death in cardiovascular disorders, coronary syndromes, and cerebral ischemia in developed countries (Wautier&Wautier, 2020).

Hemostasis and thrombosis have undergone a paradigm shift due to the development of molecular biology and cell biology. Molecular structures of blood cells and vessel walls, endothelial cell molecules, smooth muscle cell molecules, and subendothelium structures were studied after coagulation factors were discovered. As a result of gaining access to genes and gene regulation from molecules responsible for thrombotic and hemostatic processes, we reached a new level of understanding. Molecular interactions between platelets and vessel walls were detected first. Researchers have discovered the role of glycoproteins (GP IIb/IIIa) in platelet aggregation and GP Ib–IX–V complex in platelet adhesion in patients with Glanzmann thrombasthenia and Bernard–Soulier syndrome (Wautier&Wautier, 2020).

The hemorheological factors play a major role in the interaction between blood cells and vessel walls. Thrombus formation is also influenced by blood cells, in addition to plasmatic factors. Initially, platelets played a critical role in vessel occlusion and the formation of clots, but hemorheology emphasized the role of red blood cells. During the 20th century, leukocyte functions were characterized. The production of tissue factor by leukocytes during inflammation can alter endothelial cells and vessel walls.

As a result of red blood cells adhering to endothelium more tightly in sickle cell diseases, Vaso-occlusion has been described. Several diseases, including diabetes mellitus, polycythemia vera, and retinal vascular diseases, are associated with increased RBC adhesion. Additionally, polycythemia vera (myeloproliferative disorder) is frequently complicated by thrombosis. In addition, RBCs participating in this process were associated with mutations in the JAK2 kinase gene and modifications in an adhesion-related RBC molecule. Hemostasis has been studied in patients suffering from atherosclerosis, diabetes, or aging that has enabled the discovery of new molecules within the vessel and their functions. As an example, RAGE, the receptor for advanced glycation end products, has recently emerged (Wautier&Wautier, 2020).

"[Zuckerkandl] found that in the beta chain of the human and the beta chain of the horse, for example, 20 of the 146 amino acids are different; but with human and gorilla, only one is different. It is the same amount of difference, just one amino acid residue, as between ordinary humans and sickle cell anemia patients, who manufacture sickle-cell-anemia hemoglobin."

—Linus Pauling. "Medicine in a Rational Society." J. Mt. Sinai Hosp. N. Y. 36: 194–199. 1969.

2.2. RBCS AFFECT THE MOTION OF WBCS

Rolling occurs when circulating leukocytes contact the vessel wall through margination (radial migration). The interaction between white and red blood cells at bifurcations or postcapillary expansions has been associated with leukocyte margination (Pearson &Lipowsky, 2000) and also with RBC aggregates' ability to exclude WBCs from the bulk solution (Schmid et al., 1980). The use of a lattice-Boltzmann mathematical model of flow in a vessel expansion demonstrates that a coherent rouleau of RBCs is more effective than a loose association of cells at pushing a WBC to the vessel wall. Simulated forces between cells are calculated by explicitly accounting for each cell in these simulations (Chapman &Cokelet, 1998). A radial motion of the leukocyte is induced by organized groups of RBCs. RBCs aggregating depends on a number of factors, including the relative numbers of RBCs and WBCs, and fluid dynamics at the capillary inlet. Intriguingly, RBCs' shapes affect their ability to remain stacked. Compared to a train of ellipses of the same size, a rouleau of flat RBCs behaves quite differently. A flat shape can easily stack to form a rouleau, providing sufficient force to initiate WBC rolling, whereas an ellipsoidal shape tends to "roll" against one another, disrupting the stacking arrangement, and therefore providing less force to the marginating WBC.

According to other studies, retaining the Rouleau "lever" enhances the lateral force applied to the marginating leukocyte by including reversible RBC–RBC adhesion in the rouleau (Sunn& Munn, 2006). WBC margination has also been demonstrated to be strongly affected by RBC reorganization and aggregation during expansions in microfluidic devices at the scale of actual micro vessels (Munn et al., 2006). This indicates that the vessel geometry and shear-dependent flow anomalies of blood serve useful purposes, such as facilitating extravasation in postcapillaries where adhesion molecules are expressed and the vessel wall is relatively thin.

WBCs continue to be subjected to tangential and normal forces once they begin rolling on the endothelium. A microvillus could penetrate the glycocalyx due to these forces, which were estimated to be large enough (Migliorini et al., 2002). The plasma-rich zone's unique environment traps WBCs, pushes them along (by increasing tangential force), and encourages glycocalyx penetration by fluctuating normal forces once they roll on the vessel wall. When venules converge and when shear rates are high in a real vessel network, the WBC may escape the plasma-rich zone.

2.3. WBCS AFFECT THE MOTION OF RBCS

Flow resistance in the microvasculature can be greatly affected by the presence of WBCs, which are larger than RBCs (Chapman &Cokelet, 1998). Flow velocities and pressure fields were measured while simulating blood flow through channels (Sun & Munn, 2005). In addition to the hematocrit and the conduit dimensions, WBC adhesion also contributes to the flow resistance. Simulations reproduce the Fahraeus–Lindqvist effect, which predicts that blood viscosity decreases with decreasing diameter when cells suspended in pressure-driven flow are explicitly taken into account (Goldsmith et al., 1989). Since the large WBC occupies a greater fraction of the channel cross-section in the smaller tube, the resistance becomes greater in the smaller tube. In 2D simulations, the effect is somewhat exaggerated, especially in short segments of the vasculature (Sun & Munn, 2005). For example, a WBC with a diameter of 9μm blocks nearly 50% of the cross-sectional area of a 20 μm channel in 2D, but only 25% in a three-dimensional cylindrical tube.

2.4. BLOOD FLOW AFFECTS BLOOD VESSELS' PHYSIOLOGY AND VICE VERSA

Furthermore, blood cells exert mechanical forces on the vascular endothelium, influencing vessel development, remodeling, and flow regulation (Chen et al., 2001). For instance, a section of the dorsal skin of a mouse was imaged and digitized in order to study the interactions between blood cells and vessel walls. A lattice-Boltzmann simulation of flowing RBCs and WBCs was carried out by varying the initial cell configuration and aggregation of the RBCs. While passing through the vessel wall, the cells cause temporary and dramatic increases in pressure and shear stress. Angiogenesis, endothelial mitosis, and atherosclerosis can be affected by these fluctuations in force. It was also observed in simulations of deformable cells at high hematocrits in three dimensions (Dupin et al., 2007).

Every element in a complex physiological system affects the others. In the same way that blood cell dynamics affect blood vessel walls, blood vessel walls can affect blood flow as well. Contractions and dilations of smooth muscle cells are obvious examples of this. In addition, plasma extravasation and vascular permeability are modulated. Both inflammation and cancer are characterized by excessive vessel leakiness. To enable the immune response, plasma extravasation and leukocyte adhesion occur simultaneously in inflammation. Similar mechanisms have been documented in tumors, but they are poorly regulated.

2.5. BLOOD CELLS AND ENDOTHELIAL CELLS

2.5.1. Endothelium

In the inner face of vessels, there are endothelial cells (ECs). Although they share many characteristics, they are specialized in certain areas, such as the brain, kidneys, liver, skin, and coronary arteries. Depending on their location in the organ, endothelial junctions differ. Junctions are dynamic structures. Transmembrane adhesive proteins form junctional structures between endothelial cells. The junction permeation plays a major role in controlling the permeability of plasma solutes. Inflammatory areas require tightly regulated opening and closing of cell-to-cell contacts for leukocyte extravasation and infiltration (Bazzoni&Dejana, 2004). As well as regulating vessel tone, ECs also synthesize nitric oxide (NO) and prostacyclin (PGI2) (Furchgott&Zawadzki, 1980). Blood cells and endothelial cells interact through EC receptors (Gimbrone et al., 2016). The ECs synthesize the von Willebrand factor, which plays a role in hemostasis (Peyvandi et al., 2011).

2.5.2. Platelets

Coronary disease, stroke, and peripheral vascular disease were treated with antiaggregating agents during the 1970s. For example, prostaglandin G2 (PGG2) and prostaglandin H2 (PGH2) are precursors to thromboxane A2, and aspirin inhibits cyclooxygenase, which produces PGG2 and PGH2. In order to inhibit platelet activation and thrombus formation, anti-platelet drugs were developed. As far as structure and function are concerned, platelet glycoproteins (GP), GP IIb, and GP IIIa are similar to membrane proteins. These proteins are called cytoadhesins or integrins (Hynes, 1987). RGD-specific adhesion receptors are related to them. In addition, they play a role in cell-to-cell communication. Plasma membranes of platelets are predominantly composed of platelet GP IIb and IIIa (αIIbβ3). Platelet alloantigens are transported by these immunogens. Immune disorders such as allo- and autoimmune attacks can also affect them. Furthermore, Lek (Bak) is present on GP IIb, while PLA is present on GP IIIa (Kieffer et al., 1988).

The evidence of these genetic markers is significantly limited or not present in patients with Glanzmann's thrombasthenia (hereditary blood disorder) (Nurden et al., 2013). It is the GPIb-IX-V complex that is missing in Bernard-Soulier syndrome that is responsible for subendothelial adhesion to collagen (Andrews & Berndt, 2013). It is implicated in the mechanism of platelet adhesion when von Willebrand factor is absent in congenital hemorrhagic syndrome (Peyvandi et al., 2011). A growing body of research suggests that platelets may interact directly with ECs conditioned by inflammation. By expressing CD39, ECs release nitric oxide (NO) and prostacyclin (PGI2) to protect blood from thrombosis. PGI2 and NO signal into the platelet to maintain high levels of cyclic nucleotides within the cytoplasm, thereby antagonizing platelet activation pathways (Ed Rainger et al., 2015).

2.5.3. Leukocytes

Inflammation, infection, and heart disease can significantly extend leukocyte adhesion (Ed Rainger et al., 2015). In recent years, considerable attention has been paid to the leukocyte-endothelium interaction and its molecular basis (Libby, 2013). The leukocyte adhesion molecule (LeuCAM) can be studied using monoclonal antibodies. CD18 integrins (CD18a) and CD11a, CD11b, CD11c, and CD11d subunits (CD11a, CD11b, CD11c, CD11d) function along these lines in leukocytes, while on endothelial cells, VCAMs and ICAMs perform this function. The Figure 2.1 illustrates this.

Leukocytes roll, adhere, and extravasate due to the molecule interaction. Additionally, other leukocyte structures are involved in this process (Barbe et al., 2001). The fibronectin receptor is one of them. There are specific structures that are connected to endothelial cells, according to researchers (Liote et al., 1996). Intercellular adhesion molecules (ICAMs) are also expressed by these cells when they are unstimulated. Endothelial cells that have been activated can only express E-selectin and vascular cell adhesion molecules (VCAMs). As a result of cytokines, the production and expression of E-selectin and VCAM are increased.

DR, DP, and DQ can also be associated with elevated levels of ICAMs and human leukocyte antigens (HLAs). Under flow conditions, interactions between selectins and their ligands govern leukocyte adhesion to EC (Ed Rainger et al., 2015). ECs mediate firm adhesion/arrest on leukocytes by interacting with members of the CD18 family and their associated immunoglobulin ligands.

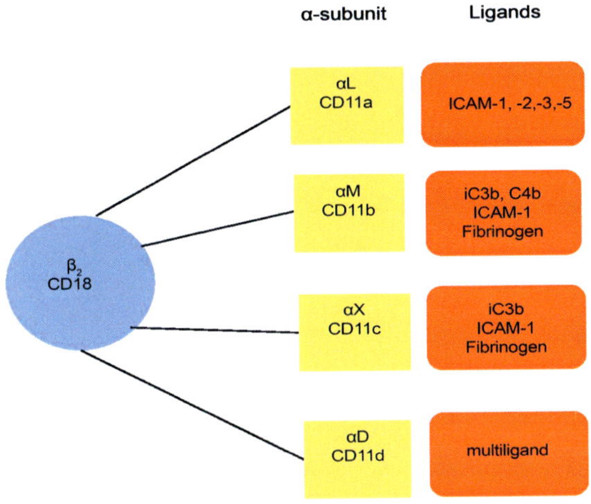

Figure 2.1. *Leukocyte integrins and their ligands.*

There are four subunits associated with the CD18 subunit (αL CD11a, αM CD11b, αX CD11c, and αD CD11d). The integrins bind to a specific ligand—intercellular adhesion molecules (ICAMs), fibrinogen, or complement components (iC3b, C4b, c3b). CD11d is a multi-ligand receptor (Source: Wautier J-L, &Wautier M-P. (2020). Cellular and molecular aspects of blood cell–endothelium interactions in vascular disorders. International Journal of Molecular Sciences, 21(15), 5315: © 2020 by the authors. Licensee MDPI, Basel, Switzerland. This article is an open access article distributed under the terms and conditions of the Creative Commons Attribution (CC BY) license: http://creativecommons.org/licenses/by/4.0/).

As a consequence of inflammatory mediators such as histamine or thrombin activating ECs, P-selectin mediates leukocyte adhesion. In this process, P-selectin is redistributed from the endothelial surface to its storage granules. In contrast, E-selectin is only expressed on the cell surface of ECs stimulated by inflammatory cytokines (e.g., IL-1 and TNF). Additionally, ICAMs act as major adhesion molecules on ECs, stopping the movement of circulating leukocytes. There are many factors that contribute to the elevation of leukocyte adhesion, including acute respiratory distress syndrome (ARDS), hypercholesterolemia, and diabetes mellitus. The vascular system is then damaged as a result. Atherosclerosis and atherothrombosis are associated with platelet-leukocyte aggregates. Hence, anti-thrombotic and anti-atherosclerotic treatments must focus on modifying leukocyte-endothelial cells (Wautier&Wautier, 2020).

Inflammatory cytokines and reactive oxygen intermediates are released by activated macrophages in atherosclerotic lesions. The breakdown of the extracellular matrix releases damage-associated molecular patterns, such as ATP and hyaluronic acid, by cardiomyocytes, tissue-resident macrophages, and fibroblasts. By activating local macrophages and neutrophils and stimulating intracellular mitogen-activated protein kinases (MAPK) and NF-κB signaling, this stimulates a proinflammatory immune response (Jones et al., 2017).

SARS-CoV-2 binding to the angiotensin-converting enzyme receptor present on endothelial cells may facilitate the interaction with leukocytes, enhancing the expression of adhesion receptors. By binding to angiotensin-converting enzyme receptors on endothelial cells, SARS-CoV-2 may enhance adhesion receptor expression on leukocytes. Thrombosis and apoptosis can result from this inflammatory reaction that produces interleukin-6 (IL6), macrophage chemoattractant protein-1 (MCP1), and tissue factor.

2.5.4. Red Blood Cells

2.5.4.1. Red Blood Cell Adhesion in Sickle Cell Anemia

There are functional abnormalities associated with sickle cell disease (SCD). RBC structure creates issues in this hematological disease. In sickling, hemoglobin S polymerizes, and red blood cells lose their deformability when the oxygen partial pressure is lowered. Those with sickle cell disease have vascular endothelial problems, according to Hebbel. Furthermore, chromium-51-labeled endothelial cells and red blood cells allowed researchers to observe the effects of this condition in an experimental setting. These issues are exacerbated during vaso-occlusive episodes (Wautier et al., 1985).

- HbS polymers are created as a result of RBCs adhering to vascular walls. Cells cannot move through capillaries as a result of this restriction. Vaso-occlusion results from this chain reaction. Postcapillary venules contain sickle cells that are trapped inside due to abnormal cell adhesion. Using erythroid adhesion molecules and endothelial cell proteins, it is possible to reduce the movement of RBCs in patients suffering from sickle cell disease. It has been reported that HbSS RBCs interact with endothelial cells of the vascular wall. A significant role is played by VLA-4, a group of integrins that form on newly formed moving reticulocytes, allowing RBCs to connect with vascular cell adhesion molecule-1 (VCAM-1) (El Nemer et al., 2007) (see the Figure 2.2 for more details). CD36 molecules are connected to HbSS reticulocytes via thrombospondin (TSP), which binds to endothelial cells. However, this does not diminish the impact of SCD in patients lacking CD36 (Lee et al., 2001).

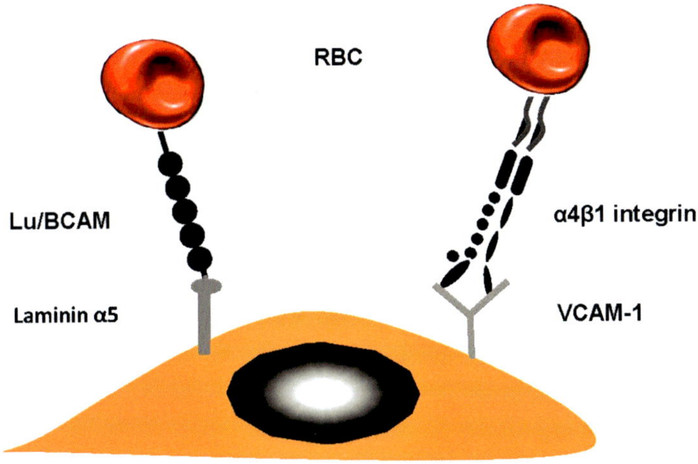

Figure 2.2.*Red blood cell (RBC) adhesion in sickle cell diseases. An important property of Lutheran Lu/BCAM is its ability to form a connection with laminin α5. Sub-endothelium occurs in this case, but apical side endothelial cells can also be affected. VLA-4 (α4β1 integrin) is a molecule produced by reticulocytes; this has a strong sticky quality.*

Source: Wautier J-L, &Wautier M-P., (2020). Cellular and molecular aspects of blood cell–endothelium interactions in vascular disorders. International Journal of Molecular Sciences, 2020; 21(15),5315: © 2020 by the authors. Licensee MDPI, Basel, Switzerland. This article is an open access article distributed under the terms and conditions of the Creative Commons Attribution (CC BY) License: http://creativecommons.org/licenses/by/4.0/).

2.5.4.2. Red Blood Cell Adhesion in Polycythemia Vera

Polycythemia vera (PV) was discovered in 1892. It was identified in a patient with thrombosis. Diagnosis of this disease can be done through a bone marrow culture, determination of red cell mass, and identifying mutations in JAK2. PV patients are at a high risk of thrombotic events, with rates ranging from 12% to 39% at the time of diagnosis in various studies. Recurrent thrombosis affects 31% of patients on the venous side and 67.6% on the arterial side. Individuals with a high level of the V617F allele mutation in JAK2 are at an increased risk of developing this condition (Vannucchi et al., 2007). The risk of developing the disease is higher in individuals with a high level of the V617F allele mutation in JAK2. This mutation was found in 34% and 45% of patients with Budd-Chiari syndrome who experienced thrombosis of hepatic, portal, or mesenteric veins, respectively (Kiladjian et al., 2008).

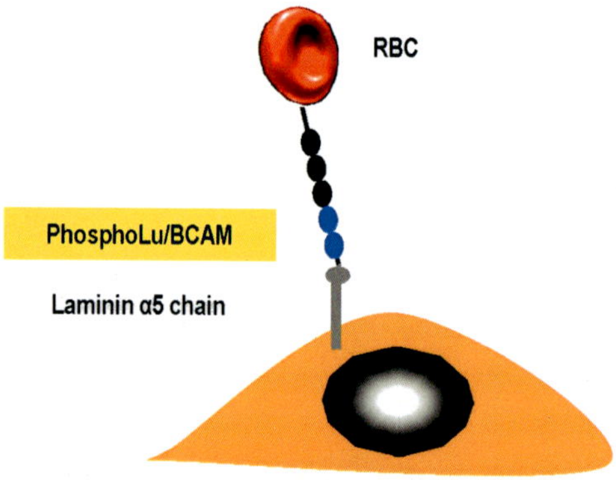

Figure 2.3.*Polycythemia vera. As a result of a JAK2 V617 F mutation, polycythemia vera patients have erythrocyte-phosphorylated Lu/BCAM.*

The subendothelium and the apical-side endothelial cells are negatively affected by this reaction. As a result of the cell adhesion to the laminin protein within a spectrin-based skeleton, the Lu/BCAM cytoplasmic domain is affected.

Source: Wautier J-L, &Wautier M-P., (2020). Cellular and molecular aspects of blood cell–endothelium interactions in vascular disorders. International Journal of Molecular Sciences, *21(15), 5315:*

Researchers have observed that PV patients exhibit increased red blood cell adhesion and have explored the molecular basis for this abnormality (Wautier et al., 2007) (Figure 2.3). There is compelling evidence that the interaction between red blood cells (RBCs) and endothelium may negatively impact PV *in vitro*. Additionally, PV RBC adhesion activates endothelial cells and enhances the expression of vascular cell adhesion molecules, promoting leukocyte adhesion. Apart from activating endothelial cells, PV RBC adhesion stimulates the expression of vascular cell adhesion molecules, which facilitates leukocyte adhesion.

Myeloproliferation may result from mutations in JAK2 found in PV patients (James et al., 2005). There is also a greater likelihood of functional abnormalities when the number of blood cells is increased. The first appearance of Lu/BCAM occurs on early orthochromatic erythroblasts during erythropoiesis after band 3 (Southcott et al., 1999). Perhaps the disease's ability to be cloned explains why it involves more than just reticulocytes in its excessive adhesiveness. It is common for PV patients to suffer from thrombotic complications and cerebrovascular problems. Budd-Chiari syndrome patients often have a post-sinusoidal obstruction in their hepatic veins beyond the liver. There may be an explanation for why thrombosis occurs systematically based on the variable response of flow.

2.5.4.3. Red Blood Cell Adhesion in Retinal-Vein Occlusion

A retinal vein occlusion (RVO) can cause vision loss in elderly people. In addition to diabetic retinopathy, it is the second most common cause of reduced vision. As a result of the anatomical classification, there are three types of occlusions: hemi-retinal-vein occlusions (HRVOs), central-retinal-vein occlusions (CRVOs), and branch-retinal-vein occlusions (BRVOs). In addition to retinal hemorrhages, venous dilation can sometimes accompany retinal swelling in a particular area, as well as RVO. For CRVO, this can occur diffusely, within the hemiretina for HRVO, or within the retinal quadrant for BRVO. Due to its location at an arteriovenous crossing, BRVO is considered the least serious form of RVO (Nentwich and Ulbig, 2015). The worst condition is central-retinal-vein occlusion (CRVO), which is poorly understood. In patients with CRVO, RBCs that express phosphatidylserine and bind to phosphatidylserine receptors are more adherent. PS RBCs bound to Annexin V inhibit endothelial cell adhesion (Wautier et al., 2011) (Figure 2.4). In contrast, CRVO is limited to 0.1%, while BRVO occurs at 0.6%. Furthermore, hypertension is associated with BRVO (Klein et al., 2000). There is a connection between strokes and central retinal artery occlusions (CRAOs). One out of every 100,000 patients is reported to suffer from this condition (Varma et al., 2013).

2.6. RBC ADHESION AND MOLECULAR ALTERATIONS IN DIABETES MELLITUS

2.6.1. Red Blood Cell Adhesion in Diabetes Mellitus

Numerous thrombotic diseases are associated with hemorheological factors. Atherosclerosis and microvascular disease are also associated with diabetes mellitus. Diabetes patients can also have abnormal erythrocytes. These issues have been linked to higher levels of viscosity in whole blood and plasma. It has also been found that this has a negative impact on erythrocyte formation. The erythrocytes of diabetic patients adhered more strongly to cultured endothelial cells after being washed. There is a correlation between red cell adhesion and the severity of diabetic vascular disease, but this does not change with age or the duration of the disease (Wautier&Wautier, 2020). There are quite a few proteins that can be glycosylated non-enzymatically by diabetics. Similarly, abnormal red cells may be caused by red cell membrane glycoproteins.

2.6.2. Advanced Glycation End Products (AGE)

In 1912, L.C. Maillard examined and explained melanoidin formation for the first time. A free amino protein group, which is primarily composed of lysine and arginine, interacts with a carbohydrate group. Amadori discovered the rearrangement of Amadori and Heyns. There is, however, still a lack of knowledge about Maillard products. The HbA1c biomarker was developed based on non-enzymatic glycosylation (glycation) of hemoglobin (Wautier&Guillausseau, 1998). Hemoglobin is glycated at two different locations. This occurs first on the valine residue of the N-terminal β chains. This is located by the α and β chains of the € amino group. Secondly, it can be found at the α chains N-termini (McDonald et al., 1978). Insulin, oral anti-diabetic therapy, and antidiabetic treatment are adjusted according to HbA1c levels. Diabetic patients have glycated RBC-band-3 membrane protein. An increased RBC adhesion observed in diabetes mellitus is due to the binding of glycated RBC band 3 to the advanced glycation end product receptor (RAGE) (Figure 2.4).

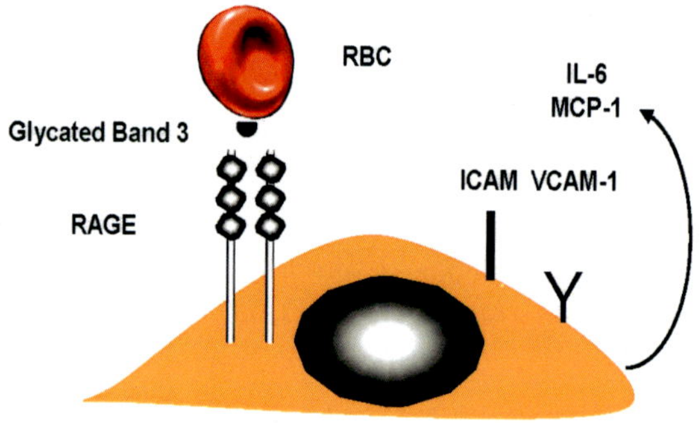

Figure 2.4.*Diabetes mellitus. Diabetes also causes glycation of the RBC-band-3 protein (AGE). When RAGE binds to ECs, it activates the cells, triggering the release of tissue factor (TF) and intercellular cell adhesion molecule-1 (ICAM-1) as well as vascular cell adhesion molecule-1 (VCAM-1). Interleukin-6 (IL-6) and vascular endothelial growth factor (VEGF) were released in response to activation.*

Source: Wautier J-L, Wautier M-P, (2020). Cellular and molecular aspects of blood cell–endothelium interactions in vascular disorders. International Journal of Molecular Sciences, 21(15), 5315:

Glycoxidation products occur due to glucose deterioration. Additionally, glyoxal and methylglyoxal (MG) contribute to AGE formation (Shipanova et al., 1997). It may be located extracellularly or intracellularly (Brownlee, 2001). There are three major AGEs: N-epsilon (carboxymethyl)lysine (CML), MG-derived AGE, and pentosidine. Dietary products contain AGEs, which are transformed within the body before being excreted by the kidneys. In addition to AGEs, glyoxalase can partially degrade them and the kidney can eliminate them, but they accumulate with age as well (Schleicher et al., 1997).

2.6.3. AGE Receptors

The original hypothesis for RAGE was that it contained two polypeptides, one lactoferrin-like (80 kDa) and another polypeptide (35 kDa). AGEs are extremely harmful to the body. RAGE can also connect a considerable amount of endogenous molecules, including aminoamyloid peptides and S100 calgranulins, and amphoterin C. RAGE, a member of the immunoglobulin (Ig) superfamily, has three Ig-like domains—an extracellular part with one variable (V) and two constants (C1 and C2), a single transmembrane domain, and one short cytosolic domain.

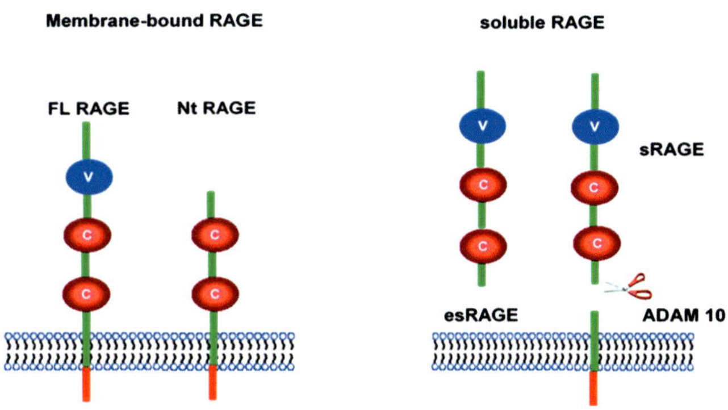

Figure 2.5.*RAGE isoforms. It is possible to produce AGE receptors as either complete molecules (Full-length, FL-RAGE) or as dominant negatives (Nt-RAGE) which is a non-efficient receptor. Soluble forms of RAGE can result from an incomplete transduction of the gene (esRAGE) or from a proteolytic reaction by ADAM 10 metalloproteinase.*

Source: Wautier J-L, Wautier M-P, (2020). Cellular and molecular aspects of blood cell–endotheliuminteractions in vascular disorders.International Journal of Molecular Sciences, 21(15), 5315: © 2020 by the authors. Licensee MDPI, Basel, Switzerland. This article is an open access article distributed under the terms and conditions of the Creative Commons Attribution (CC BY) license: http://creativecommons.org/licenses/by/4.0/.

The polymorphism of RAGE has been examined by researchers (Prevost et al., 2005). There are up to 20 different forms of RAGE transcripts when they are separated (Hudson et al., 2008). It is important to note, however, that endothelial cells produce only three types of RAGE. These include N-truncated (Nt-RAGE), full length (FL-RAGE), and endogenous secretory (esRAGE). By using FL-RAGE proteolysis, soluble RAGE (sRAGE) can be produced (Figure 2.5). Throughout the body, RAGE occurs at minimal levels.

In addition, methylglyoxal human serum albumin (MG-HSA) and N-(carboxymethyl) lysine human serum albumin (CML-HSA) interact with the receptor and affect the expression of RAGE isoforms. This type of cell contains three forms of RAGE, which are affected by MG-HSA. FL- and Nt-RAGE isoforms are the only ones that carry CML-HAS. MG-HSA and CML-HSA both contribute to the expression of RAGE. In spite of this, MG-HSA increases esRAGE expression, possibly implicating a negative feedback loop in which soluble RAGE acts as a decoy intercepting ligands' interaction with cell-surface RAGE. Cellular activation mediated by RAGE could be restricted by this (Grossin et al., 2009). RAGE ligands could damage the vascular bed when RAGE isoform expression affects it. Diseases can be accelerated by FL and ligand interactions. Given that they lack cellular signaling domains, they can also restrict positive actions.

2.6.4. Age and Rage

For many years, researchers have studied the relationship between glycated proteins and vascular disease. In addition to animal studies, glycated proteins have been studied in relation to nephropathy. The anti-RAGE antibody was found to limit the increased permeability of vascular walls in rats suffering from diabetes. Inflammation and renal failure are also associated with AGE. RAGE interacts with N-epsilon (carboxymethyl) lysine (CML)-adducts, affecting signal transduction pathways.

Genes are affected by the receptor RAGE. The RAGE cytoplasmic domain (ctRAGE) interacts both *in vitro* and *in vivo* with diaphanous-1 (DIAPH1). A number of key consequences occur when ctRAGE interacts with DIAPH1, including modulation of actin cytoskeleton and cell migration, activation of Rho GTPases, and production of reactive oxygen intermediates. Some types of RAGE lack the intracellular domain, which is able to attach to ligands. In addition, it has a "dominantly negative" impact on CML-adducts. CML binding to RAGE activates NF-κB; however, DN-RAGE-transfected cells do not really exhibit this effect.

Reactive oxygen intermediates (ROI) are generated by the interaction of AGE on the surface of red blood cells or AGE immune-isolated from the blood of patients with diabetes or renal failure with FMIG (Refer to the Figure 2.6) (Wautier& Schmidt, 2004).

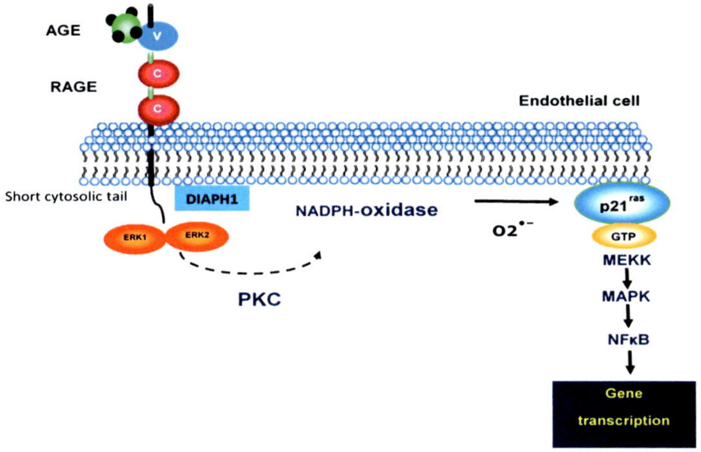

Figure 2.6.*Binding of AGE to the endothelial cell receptor, RAGE. After this, the activation of NADPH oxidase and the formation of reactive oxygen intermediates are caused by a series of reactions involving diaphanous-1 (DIAPH 1), ERK1, ERK2, and protein kinase C. Through a pathway including the MAP kinase, NF-κB becomes activated and induces gene transcription.*

Source: Wautier J-L, &Wautier M-P, (2020).Cellular and molecular aspects of bloodcell–endothelium interactions in vascular disorders. International Journal of Molecular Sciences, 21(15), 5315: © 2020 by the authors. Licensee MDPI, Basel, Switzerland. This article is an open access article distributed under the terms and conditions of the Creative Commons Attribution (CC BY) license: http://creative-commons.org/licenses/by/4.0/.

As a result of RAGE, both the substrate and the subcellular location of ERK can be altered. Also, it can interact with intracellular signals through NF-κB. As a result, researchers are trying to determine which molecular mechanism is responsible for its effects. It is also important to consider the ROI as it is crucial to the cellular ligation of RAGE via AGE. A ROI was observed following the activation of NADPH oxidase by the AGE-RAGE interaction. Gene expression in a specific cellular microenvironment was then affected by this chain of events. RAGE analogs TTP488 (azeliragon) and recombinant rRAGE also protect against the damaging effects of RAGE when they are treated with anti-RAGE antibodies (refer to the Figure 2.7).

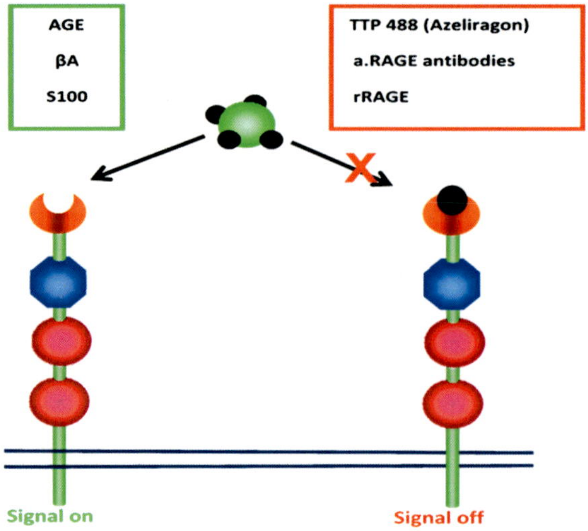

Figure 2.7. *Blockade of RAGE activation. RAGE molecule has an extracellular part with two constant domains (pink circles) and one variable domain (blue octagon). By using anti-RAGE antibodies or synthetic peptide TTP488, AGE, amyloid protein, and S100 calgranulins can be inhibited from binding to the RAGE region (brown). In contrast, recombinant RAGE binds AGE ligands that compete against cellular RAGE. Endothelial cells were unable to transduce signals due to the blockade.*

Source: Wautier J-L, & Wautier M-P, (2020). Cellular and molecular aspects of blood cell–endothelium interactions in vasculardisorders. International Journal of Molecular Sciences, 21(15):5315: © 2020 by the authors. Licensee MDPI, Basel, Switzerland. This article is an open access article distributed under the terms and conditions of the Creative Commons Attribution (CC BY) license: http://creative-commons.org/licenses/by/4.0/).

2.7. INTERACTION OF RBCS AND ENDOTHELIAL (ECS)

This unusual aggregation of RBCs and their adhesiveness to other RBCs and to ECs have been linked to various vascular disorders, such as SCD, Diabetes, and hypertension (Setty et al., 2002). Patients with type 2 diabetes mellitus (T2DM) were shown to be susceptible to non-absorbing macromolecules mediating RBC adhesion to ECs (Kaliyaperumal et al., 2019). Macromolecular depletion has been proposed as an effective mechanism for enhancing adhesion between cells and surfaces when studying nonspecific forces, such as attractive and repulsive electrostatic forces (Neu & Meiselman, 2006). Microvascular blood flow behaves differently from macrovascular blood flow due to an additional resistance on the luminal surface of the smaller blood vessels because of the endothelial glycocalyx layer (Liu & Yang, 2009). It is the glycocalyx on the surfaces of both RBCs and ECs that influences this dynamic interaction between RBCs and ECs.

Due to the anionic properties of endothelial glycocalyx, these layers facilitate "cushion" functions by preventing negatively charged RBCs from coming into contact (Liu & Yang, 2009). As a result of metabolic syndrome, inflammation, and excessive sodium consumption, the endothelial glycocalyx can be damaged (Oberleithner et al., 2011). Due to increased oxidative stress and aging of the RBC, the glycocalyx gets damaged (Neu et al., 2003). In the dynamic interaction between these cells, damaged EC glycocalyx leads to the shedding of RBC glycocalyx (Oberleithner, 2013). The interaction between the RBC and EC further enhances the defect when it occurs in either the RBC or EC glycocalyx. Considering that the glycocalyx prevents adhesion (Liu & Yang, 2009), such a defect could increase RBC adhesiveness to ECs. However, *in vivo* testing has not yet been performed on this hypothesis.

2.8. INTERACTION OF RBCS AND PLATELETS

Hydrodynamic interactions are another type of interplay described between cells. The concentration of platelets near the vessel wall is highest in a process called platelet margination or lateral drift (Vahidkhah et al., 2014). RBCs seem to be involved in this interaction (Carboni, 2017). There is either a crossing or a turning interaction between a platelet and an RBC. Crossing interactions occur when the platelet rolls over the RBC, slightly deforming the flexible RBC. During the turning type, the platelet approaches the RBC but then reverses back without making direct contact with it. The initial lateral separation between a platelet and an RBC determines whether they cross or turn (Vahidkhah et al., 2013). This lateral separation is the only factor determining platelet-RBC interaction in the midflow of the vessel. The movement is affected, however, if the interaction occurs closer to the vessel wall (Bächer et al., 2018). A platelet located further from the vessel wall than the RBC will have a higher likelihood of undergoing a turning interaction trajectory, while a platelet located closer to the vessel wall will have a higher probability of undergoing a crossing interaction (Vahidkhah et al., 2013).

As a result of these processes, platelets are continuously driven away from the RBC-rich area in the center of the vessel and into the plasma layer close to the vessel wall (Vahidkhah et al., 2013). According to Tokarev et al., platelets are forced against the vessel wall by rebounding collision with an outrunning RBC as a result of heart rate, hematocrit level, and RBC size (Fitzgibbon et al., 2015). It may also be the result of these mechanisms that elevated hematocrit enhances platelet accumulation and adhesion to the vascular wall. A high hematocrit also increases platelet interaction with thrombi, resulting in thrombi formation and platelet accumulation (Walton et al., 2017). RBC rigidity also plays a role in thrombosis susceptibility: rigid RBCs increase platelet marginalization. Blood flow dynamics and blood cell transport are affected by abnormal RBC and platelet properties and biorheology in T2DM. Interestingly, less deformable T2DM RBCs reduce heterogeneous collisions with the highest near-wall accumulation of T2DM platelets (Chang et al., 2018).

2.9. INTERACTION OF RBCS AND WBCS

Flowing WBCs must migrate in a radial direction to roll against the endothelium (Munn &Dupin, 2008). As a result of RBC's ability to aggregate and exclude WBC from the bulk solution, this phenomenon is called margination. A loosely associated group of cells is less effective at pushing WBC against the vessel wall than the rouleau formation of RBC.

A number of factors affect WBC margination, including the hematocrit, the flow rate, and the RBC as well as the deformability of the WB (Fedosov&Gompper, 2014). Due to their large size, WBCs reverse the Fahraeus-Lindqvist effect when bounced against the endothelium by the RBC (Munn &Dupin, 2008). The pathology of the microvasculature is exacerbated by adherent WBC in conditions such as SCD (Finnegan et al., 2007).

2.10. INTERACTION OF RBCS AND MACROPHAGES

There has been much discussion about the interaction between RBCs and macrophages. Both erythropoiesis and erythrophagocytosis are directly regulated by macrophages.

In the bone marrow niche, macrophages provide signals that cause RBC progenitors to differentiate and multiply. Splenic and liver macrophages remove defective RBCs from the circulation by filtering, repairing, and/or removing them from the circulation (Klei et al., 2017).

Regarding these two fundamental stages, specific and detailed reviews of health (Badior& Casey, 2018) and diseases (Flegel, 2015) are warranted in the future. We do not discuss interactions that occur during maturation and clearance of RBCs in this section because the focus is on those that occur during the life of an RBC.

2.11. RBCS INTERACTION WITH PATHOGENS

There is more to red blood cells than transporting oxygen and carbon dioxide (Morera&MacKenzie, 2011) and governing blood flow distribution (Minasyan, 2018). During phagocytosis, bacteria are cleared from a dynamic compartment such as the bloodstream. RBCs are consequently attracted to bacteria moving with the blood flow because they are triboelectrically charged (Minasyan, 2016). During this contact, oxygen is released from oxyhemoglobin and reaches the surface of the RBC, causing it to die. Ultimately, the bacteria are washed from the RBC surface and digested by the liver or spleen as a result of this diminishing triboelectric charge (Minasyan, 2018). The effect of free hemoglobin during homeostasis may confer antimicrobial protection, but during disease, such as severe sepsis, it has the opposite effect, often resulting in an increase in mortality (Anderson et al., 2018).

Glycophorins can bind many pathogens, including reovirus, *influenza C virus*, Sendai virus, *Mycoplasma pneumoniae, Escherichia coli,* and*Ureaplasmaurealyticum,* acting as chaperones to avoid important tissues and facilitate clearance by macrophages in the spleen (Baum et al., 2002). However, *Plasmodium falciparum* parasites adhere to and invade RBCs via the same glycophorins (Anderson et al., 2018). Malarial parasites are capable of multiplying and evolving inside RBCs, digesting hemoglobin and converting it into crystals known as hemozoin (Parroche et al., 2007). Both HIV-1 virions and Zika viruses do the same thing, binding to RBCs (Beck et al., 2009) and hiding inside them to enhance viral infectivity (Stone et al., 2017). RBCs can be beneficial or harmful to the host, depending on whether pathogens bind to them or internalize them (Anderson et al., 2018).

2.12. INFLAMMATION

As a response to injury or infection, the organism mounts an inflammation, which is a protective response from cellular and humoral elements. In addition to removing the originating insult, it also promotes the restoration of normal organ function and homeostasis in multicellular organisms (Medzhitov, 2008).

During inflammation, immune cells carry noxious agents ready for release into the injured environment. Neutrophils, a subset of myeloid leukocytes, are the main players during this initial phase of inflammation. During inflammatory conditions, neutrophils can rapidly increase in number in circulation, and their half-lives are short (6–8 hours). They account for about 10% of leukocytes in murine blood, and up to 70% in human blood. The importance of neutrophils for the defense against infecting pathogens is highlighted by the moderate to severe susceptibility to infections observed in animals or humans with mutations in genes involved in neutrophil trafficking (such as leukocyte adhesion deficiency types I, II and III, which affect integrins, selectin ligands and activators of integrins, respectively) (Etzioni, 2009), neutrophil generation or survival (Palmblad and Papadaki, 2008) or microbicidal capacity.

Platelets also play an important role in the initial response to injury. Human megakaryocytes release anucleated fragments with a short half-life (about 4 days). Their granules contain numerous substances that contribute to different aspects of hemostasis and immunity (including nucleotides, secretagogues, adhesion receptors, cytokines, mitogenic factors, and coagulation factors) (Ren et al., 2008). Due to the primary role of platelet function in hemostasis, bleeding disorders typically result from defects in platelet genes.

The immune response relies heavily on platelets; however, their importance is increasingly recognized (Semple & Freedman, 2010). Due to their ability to bind other blood components, platelets can influence the adaptive immune response for various reasons: binding to lymphocytes controls their homing to lymph nodes, while binding to monocytes and neutrophils enhances their function as well as their recruitment to extravascular spaces. Malaria parasites are directly eliminated by platelets by interacting with infected RBCs (McMorran et al., 2009).

By releasing microbicidal agents or phagocytosing, neutrophils and platelets allow the elimination of the originating stimuli and contribute to the repair of damaged tissues (e.g., by secreting growth factors). There is increasing evidence that neutrophils and platelets can enhance the protective inflammatory response beyond their individual functions (Zarbock et al., 2007). Untimely activation of this response within blood vessels can alter the integrity of the vessels and damage the tissues they irrigate; this phenomenon, however, is a double-edged sword. In both acute (e.g., sepsis, acute organ injury, or pain crises in patients with sickle cell disease) and chronic (e.g., atherosclerosis) settings, these processes shape the inflammatory response.

2.13. INITIATION OF INFLAMMATION AND NEUTROPHIL RECRUITMENT

An inflammatory response occurs in response to antigens, foreign organisms (such as bacteria or fungi), or trauma. When tissue-resident sentinel cells (e.g., macrophages or dendritic cells) detect danger signals, they produce pro-inflammatory cytokines such as interferons and interleukins. As a result, these cytokines activate the endothelium lining the vessels that irrigate the affected tissues, which results in adhesion receptors, as well as more cytokines and chemokines that eventually allow circulating leukocytes to be recruited (Wagner

&Frenette, 2008). It is believed that the first leukocyte subset to accumulate after injury is a subset of monocytes identified by intravital imaging (Auffray et al., 2007).

As soon as an injury occurs, neutrophils quickly and massively accumulate at the site. Neutrophils are constitutively equipped with receptors capable of mediating interactions with inflamed microvasculature, unlike other leukocyte subsets involved later in inflammation. As a result, neutrophils and endothelial cells initiate contact when adhesive molecules are induced on their surfaces. In the endothelium and on neutrophils, P and E-selectins mediate these initial contacts (Zarbock& Ley, 2009) (Figure 2.8). The Weibel-Palade bodies of endothelial cells contain P-selectin, which is also found in platelets' alpha granules, and is displayed rapidly on the cell surface following activation, thus assisting in the recruitment of leukocytes. To maximize leukocyte recruitment, E-selectin requires transcriptional activation and appears along with counter-integrins (mainly ICAM-1 and VCAM-1) on the endothelium. Selectin-mediated interactions cause rolling-like motions at physiological shear forces, which expose leukocytes to chemokines on the endothelial surface because of the optimal on-off binding rates (McEver& Zhu, 2010).

P-selectin engages P-selectin glycoprotein ligand-1 (PSGL-1) to capture or tether circulating neutrophils. Through the engagement of a variety of glycoconjugate ligands on the surface of neutrophils, E-selectin mediates slow rolling motion and promotes leukocyte arrest (Zarbock et al., 2007). E-selectin ligand-1 (ESL-1) and PSGL-1 account for the majority of E-selectin ligand activity on mouse neutrophils, while the full repertoire of these ligands remains to be characterized (Hidalgo et al., 2007). Neutrophils have been found to express CD43, a ligand on inflammatory T lymphocytes, as well as lipids and O-glycans, but their functions and identities are still unclear (Yago et al., 2010). Additionally, both P- and E-selectins contribute to the activation of rolling neutrophils beyond their role in leukocyte tethering and rolling. As well as promoting integrin activation and microdomain formation, selectins also trigger inflammation mediated by these cells, which is an important and underappreciated aspect of inflammation. P-selectin binds to PSGL-1, resulting in the initiation of tyrosine phosphorylation and Src family kinase activation, resulting in Mac-1 (CD11bCD18) activation on neutrophils (Wang et al., 2007). Conversely, PSGL-1 or CD44 engagement by E-selectin promotes the partial activation of LFA-1, another beta 2 integrin present on neutrophils, by sequentially activating SFK, Syk, Bruton, and p38 MAP kinases (Yago et al., 2010). In turn, activated LFA-1 engages endothelial ICAM-1, reduces rolling velocity, and promotes leukocyte recruitment and arrest.

A different receptor for E-selectin, ESL-1, is also involved in activating the integrin Mac-1 within specific microdomains of adherent neutrophils, which also involves SFK signaling. This suggests that there is a functionally important link between integrin activation and selectin ligands, which is supported by a complex and rich signaling crosstalk between them. The full activation of rolling neutrophils is mediated by chemokine-triggered signaling following selectin engagement. It is the chemokine CXCL1/KC (IL-8 in humans) that induces these events in neutrophils through its receptor CXCR2. The G alpha-I mediated signaling is also involved in the actions of bacterial peptides, chemoattractants, and several CXC-chemokines on neutrophils.

Firm adhesion was considered the last event in the classical view of leukocyte recruitment. According to *in vivo* imaging studies, however, most leukocytes do not immediately extravasate but rather crawl along the endothelial surface, a process mediated by the integrin Mac-1 (Chiang et al., 2007). Activated neutrophils spend longer inside blood vessels during this crawling process, and therefore spend more time inflicting vascular injury if they become abnormally activated during that time.

2.14. PLATELET RECRUITMENT AND HETEROTYPIC AGGREGATES

Platelet interactions with extracellular components or other cells play an important role in hemostasis, immunity, and inflammation. In addition to collagen, von Willebrand factor (VWf), fibrin(ogen), or fibronectin, platelet adhesion is also mediated by cellular receptors such as PSGL-1 and Mac-1, depending on the surface (Figure 2.8).

2.14.1. Platelet-Platelet Interaction

ADP, thrombin, thromboxane A2, and collagen are exogenous agonists that activate the platelet glycoprotein (GP) GPIIb/IIIa receptor (integrin alphaIIbbeta3, CD41/CD61), the most abundant protein on platelets. This integrin recognizes fibrinogen (from plasma or platelets) as its ligand, and its dimeric structure facilitates platelet aggregation. The aggregation of platelets occurs through the binding of vWf to two platelet receptors, GPIb and GPIIb/IIIa, without any additional exogenous stimuli (Nesbitt et al., 2002). Platelet interaction is also mediated by fibronectin, vitronectin, and CD40 ligands (Jackson, 2007). Under low shear conditions, fibrinogen is the dominant ligand supporting platelet aggregation, even though all ligands bind to activated GPIIb/IIIa. Thrombin converts fibrinogen into fibrin in the final phase of thrombus formation, thereby stabilizing and anchoring platelet aggregates. Platelets and fibrinogen are also formed by collagen (Patel et al., 2003). Platelet deposition and lateral cohesion are prevented by blocking GPIIb/IIIa, and platelet-collagen adhesion is altered by blocking GPIIb/IIIa (Patel et al., 2003).

2.14.2. Platelet-Vessel Wall Interactions

Intravital microscopy analysis showed that activated platelets roll against the vessel wall when activated with thrombin, collagen, or calcium ionophores. Likewise, platelet rolling is modulated by endothelium-expressed P-selectin and vWf, as well as their platelet counterreceptors, PSGL-1 and GPIbalpha (Andre et al., 2000) (Figure 2.8). It takes only minutes for platelets to disengage from vessel walls without additional stimulation. Disintegrin-like metalloprotease ADAMTS13 (a disintegrin-like metalloprotease with 13 thrombospondin repeats) cleaves vWf complexes to release them (Motto et al., 2005).

GPIb/IX/V mediates the initial interaction between platelets and collagen and vWf components of the exposed subendothelial membrane in the presence of vascular damage (Yip et al., 2005). The recognition of collagen via GPVI contributes to platelet aggregation and adhesion to the vessel wall (Massberg et al., 2004). Fibrinogen, vWf, and fibronectin are then engaged by GPIIb/IIIa to mediate binding to the vascular wall. A GPIb binding or a G protein-coupled receptor triggers the activation of platelet GPIIb/IIIa due to its prothrombotic role.

2.14.3. Platelet-Leukocyte Interactions

A variety of leukocytes, including granulocytes, monocytes, lymphocytes, and even blood progenitor cells, interact with platelets and platelet-derived microparticles (Figure 2.8). After P-selectin is released from platelet alpha-granules, activated platelets are avidly bound by leukocytes since they have a much higher density than endothelium. As with endothelial cells, leukocytes can migrate on platelets and roll on them, facilitating leukocyte recruitment to inflamed or injured vessels.

Depending on the cell subset involved, platelet interactions with leukocytes have different consequences. As monocytes bind to platelets, they are able to activate and recruit to injury sites (Huo et al., 2003), as well as to differentiate into macrophages and transform into foam cells (Stellos et al., 2008), thus contributing to atherogenic processes (Langer et al., 2010).

In the circulating lymphocyte population, platelet-lymphocyte aggregates make up about 3%. Activation of platelets increases platelet-T lymphocyte aggregates proportionally to T cell subsets' ability to bind P-selectin. The formation of heterotypic aggregates requires P-selectin, as well as CD41, CD40L, and CD11b (Li et al., 2006). By recognizing carbohydrate ligands present on high endothelial venules, P-selectin facilitates the entry of naive lymphocytes into peripheral lymph nodes, although this is mostly observed in the absence of L-selectin.

The binding of platelets to hematopoietic and endothelial progenitor cells (HPC) has also been demonstrated in several studies. The implications of these reports may have clinical significance since HPC can regenerate in atherosclerosis or ischemic heart disease. Through the presentation of P-selectin and GPIIb, and the deposition of chemokines, in particular, CXCL12, adhered platelets can recruit circulating progenitors to areas of injury. In addition, platelets may be more likely to recruit to inflamed tissues or to home to the bone marrow in transplantation settings because of their additional receptors (Janowska-Wieczorek et al., 2001). By modulating the differentiation program of progenitor and monocyte subsets, platelets may modulate inflammation beyond leukocyte trafficking.

2.15. HETEROTYPIC INTERACTIONS AT THE SITE OF INFLAMMATION

2.15.1. Leukocyte Polarization

It has been well described that leukocyte polarization involves not only rapid (in a matter of seconds) and marked changes in morphology, but also the segregation of receptors on the cell surface and signals and structural molecules in the cytoplasm (Barreiro et al., 2007) (Figure 2.8). As leukocytes sense and respond to gradients of chemoattractants, leukocyte polarization is crucial for direct migration during extravasation and tissue migration (Kinashi, 2005). As a result of chemoattractants or chemokines engaging G protein-coupled receptors, polarization occurs, resulting in an elongated, non-adhesive structure at the rear end of the cell (with respect to the direction of movement) referred to as the trailing edge or uropod, and a flat area at the front with protruding membranes referred to as the leading edge. The membrane protrusion force is generated by the polymerization of F-actin at the leading edge, enabled by specific accumulation and activation of Rho GTPases, particularly Cdc42, Rap1 and Rac (Filippi et al., 2007), along with proteins that control their activity (Nishikimi et al., 2009), such as the guanine nucleotide exchange factor DOCK2 or phosphoinositide kinases. The formation of the uropod, however, depends on Rho Rock signaling (Lee et al., 2004) and phosphatase and tensin homologs (PTEN) (Heit et al., 2008) during leukocyte polarization. The cytoplasmic segregation of transmembrane receptors is paralleled by the transmembrane receptors accumulating at the cell surface: active integrins and chemokine receptors are concentrated at the leading edge, while ligands for integrins and selectins (such as PSGL-1, CD43, CD44, L-selectin, ICAM-1 and 3) are mainly found at the uropod (Chiang et al., 2007).

It is important to note that polarization can be induced by additional factors besides chemokines. CD44 engagement by E-selectin can cause rolling neutrophils to cluster L-selectin or PSGL-1 receptors (Hidalgo et al., 2007). There is a possibility that these structures constitute "primordial" uropods and that leukocyte polarization is specified early during leukocyte recruitment. The polarization of leukocytes may be particularly relevant to heterotypic aggregate formation because the receptors that mediate these intercellular interactions are segregated and accumulated in small microdomains with a higher density, which makes them more efficient at binding ligands under flow conditions.

2.15.2. Heterotypic Interactions Mediated by Leukocyte Microdomains

Inflamed microvasculature neutrophils can act as anchors for circulating cells, which represent a variation of the scenario outlined previously, where platelets are recruited to sites of vascular damage, allowing circulating leukocytes to be rolled and arrested (Figure 2.8).

Using high-speed multichannel intravital microscopy, a study demonstrated that neutrophils that were firmly attached to the inflamed endothelium or crawling actively captured circulating platelets on a regular basis (Hidalgo et al., 2009). Detailed microscopic analysis revealed that these interactions mostly occurred at the leading edge of neutrophils, where pseudopodia are actively forming. In this microdomain, platelet capture might be mediated by one or more receptors that are specifically expressed and active. TNF alpha-induced inflammation also causes interactions with the rear side of crawling leukocytes, but these interactions represent a minor fraction (Hidalgo et al., 2009). The leading edge and uropod played a similar role in mediating interactions among leukocytes and platelets in non-inflamed vessels (Hidalgo et al., 2009). Consequently, neutrophil-platelet aggregates form at inflammatory sites differently from those found in circulating blood because they are initiated by recruited leukocytes and involve specific microdomains on the polarized, adherent neutrophils.

Besides neutrophil interactions with platelets, *in vitro* and *in vivo* imaging has shown that neutrophils also interact frequently with circulating RBCs under inflammatory conditions (Hidalgo et al., 2009). These interactions, however, are exclusively mediated by the leading edge (Hidalgo et al., 2009), implying that platelets or RBCs are interacting through partly different receptors. Furthermore, adhering neutrophils have been shown to interact with other circulating leukocytes, which enhances the recruitment of leukocytes. L-selectin and PSGL-1 bind to the uropod of adherent cells and mediate these responses (Sperandio et al., 2003).

2.15.3. Neutrophil-Platelet Interactions and Inflammatory Injury

In vitro studies have shown that platelet P-selectin and leukocyte PSGL-1 interact to tether unpolarized neutrophils to platelet-coated surfaces (Wang et al., 2005). Following initial contact, the platelet glycoprotein GPIbalpha is engaged by the leukocyte integrin Mac-1 (Wang et al., 2005). According to an analysis of mice deficient in Mac-1, P-selectin, or PSGL-1, similar receptor-ligand pairs interact with polarized, adherent neutrophils and circulating platelets. These mice lack Mac-1, resulting in a marked reduction in leading edge interactions, while uropod interactions remain unaffected (Hidalgo et al., 2009). In contrast, when P-selectin or PSGL-1 are absent, uropod interactions are specifically reduced. In these interactions, GPIbalpha or other platelet ligands may be the physiologically relevant ligands for Mac-1 (Figure 2.8).

In the context of heterotypic interactions, it is intriguing how platelet receptors are segregated into different microdomains in adherent neutrophils (Mac-1 at the leading edge, PSGL-1 at the uropod); the functional implications of platelet interactions with each domain for neutrophil activation are also intriguing.

2.15.3.1. How Do These Interactions Affect Pathophysiology?

By phagocytosing activated platelets, Maugeri and colleagues propose that neutrophils protect the cardiovascular system by removing them from circulation (Maugeri et al., 2009). Neutrophil-platelet interactions have been studied in several animal models for further elucidation of their pathology. Antibody-induced transfusion-related acute lung injury (TRALI) in mice, which is a leading cause of transfusion-related mortality and morbidity (Looney et al., 2010), causes platelets to rapidly rosette around adherent neutrophils in a Mac-1 and E-selectin-dependent manner (Hidalgo et al., 2009). It was shown that adherent neutrophils produced radical oxygen species (ROS) within minutes of challenging, and these in turn caused endothelial injury, which was prevented by ROS-scavenging molecules, which prevented endothelial permeability and lung injury (Hidalgo et al., 2009). Lung injury in mice treated with antibodies is significantly reduced by inhibiting E-selectin or depleting neutrophils or platelets (Hidalgo et al., 2009). In this model, aspirin inhibition of platelet cyclooxygenases also prevents neutrophil-mediated injury (Looney et al.,

2009), suggesting that activating platelets and neutrophils need to be physically interacting in order for vascular injury to occur. This finding is consistent with the previously described ability of activated platelets to trigger neutrophil oxidative bursts. A blockade of Mac-1 did not appear to protect mice from injury in these studies, while inhibition of P-selectin did not (Looney et al., 2009).

The pulmonary microvasculature and the circulation were also markedly increased in an acid-induced lung injury model. Endothelial damage and lung injury were induced by these interactions mediated by P-selectin (Zarbock et al., 2006). The model demonstrates how different microdomains and intercellular adhesive pathways can contribute to vascular injury, regardless of whether platelets bind unpolarized neutrophils in the circulation.

Sepsis, which kills about half a million people in the United States alone, is another example of how these interactions contribute to vascular injury. Lipopolysaccharide (LPS) and other bacterial products activate platelets and bind neutrophils to the liver sinusoids in a TLR4-dependent manner, as determined by Kubes and colleagues (Clarke et al., 2007). Postsinusoidal liver veins were inflamed and died as a result of platelet trapping by neutrophils. P-selectin did not appear to play a role in these interactions as it did in TRALI (Clarke et al., 2007). This study did not measure ROS production but exacerbated neutrophil activation was demonstrated by the production of neutrophil extracellular traps or NETs, webs of nuclear DNA containing granules and nuclear proteins that efficiently capture and kill bacteria (Brinkmann et al., 2004).

As a result of platelet engagement by neutrophils, toxic substances are released into the vascular milieu, which then damages organs. Neutrophils are likely to be activated in this context through "outside-in" signaling events triggered by engaged integrins, which are mediated by adaptor proteins like Vav-1 and 3 or Kindlin-3 (Moser et al., 2009). Therapeutic drugs could be developed targeting these and other proteins involved in neutrophil activation at sites of vascular injury.

2.15.4. Neutrophil-Erythrocyte Interactions

Interactions between adherent neutrophils and RBCs were detected under controlled *in vitro* conditions and were shown to be sensitive to the applied shear rates, with very few interactions at higher shear rates (Hidalgo et al., 2009). This reduced capture rate may be due to the higher volume of these cells compared to platelets (and concomitant drag force under flow conditions) as well as a different affinity for the ligand(s) that mediate binding, suggesting that neutrophil-RBC interactions have a limited physiological function. These interactions are absolutely dependent on the leukocyte integrin Mac-1, which was demonstrated using blocking antibodies and gene-targeted mice (Hidalgo et al., 2009). SFK signaling enhances the activity of Mac-1 at the leading edge of leukocytes through the engagement of neutrophil ESL-1 by E-selectin (Hidalgo et al., 2009) (Figure 2.8). Despite the fact that the identity of the ligands on the surface of RBCs that might bind Mac-1 is unknown, ICAM-4 and the complement protein C3 appear to be potential ligands (Goel & Diamond, 2002). Neutrophil-RBC interactions are indeed significantly reduced in mice deficient in C3 (Hidalgo et al., 2009). These interactions are thought to remove opsonized and phagocytosed aged or damaged RBCs.

A murine model of sickle-cell Vaso-occlusion demonstrates the pathophysiological consequences of these interactions. Several ischemic end-organ damage factors, such as pulmonary hypertension, renal failure, and cerebrovascular injury, contribute to the morbidity and mortality of sickle cell disease (SCD) (Frenette et al., 2004).

There are several mouse models of SCD that faithfully replicate the major characteristics of the human disease, such as anemia and chronic inflammation, by trapping sickled RBCs within the vasculature, causing endothelial activation, Vaso occlusion, and organ damage (Frenette&Atweh, 2007). Based on their study of sickle-shaped RBCs, Frenette and colleagues observed that neutrophils recruited to inflamed venules of the cremaster muscle interact with sickle-shaped RBCs at high frequency and that this phenomenon caused Vaso-occlusive episodes (Turhan et al., 2002).

As a result, the formation of neutrophil-RBC aggregates was completely abolished when E-selectin was absent (Hidalgo et al., 2009). In addition to using inhibitory antibodies (Hidalgo et al., 2009) or genetically deleting Mac-1, these interactions in sickle animals were demonstrated to be dependent on this receptor. Evidence from SCD patients suggests a role for this complement protein in the pathogenesis of Vaso occlusion.

Furthermore, elevated C3 levels in deoxygenated RBCs from SCD patients suggest that this complement protein may play a role in Vaso-occlusion initiation. Identifying pathways involved in acute inflammation in healthy animals identifies attractive therapeutic targets for chronic inflammation in a model of chronic inflammation.

It has been demonstrated that synthetic pan-selectin inhibitors can reduce intravascular interactions, improving blood flow and survival in challenged "sickle" mice. In several studies, heterotypic interactions have also been proposed to cause deep vein thrombosis pain and vascular occlusion (Goel & Diamond, 2002).

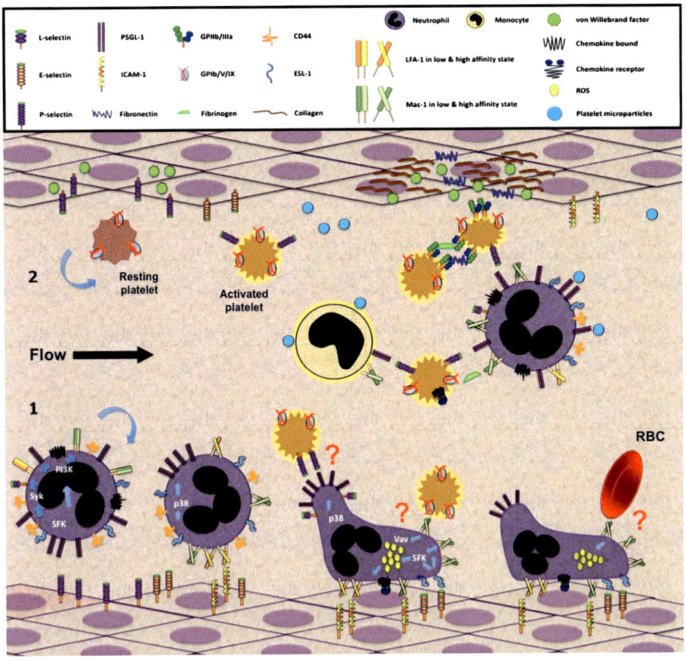

Figure 2.8.*(1) Leukocyte interactions and microdomain formation. As leukocytes roll and tether to PSGL-1, CD44 and ESL-1, they are bound by P- and E-selectins. LFA-1 and Mac-1 integrins are partially activated by these interactions, as well as Src kinase activity. The slow rolling is mediated by CD44 and ESL-1, which leads to the activation of neutrophils, which causes molecules to be polarized to the leading edge of the uropod. Different components of the blood interact with each other because intravascular leukocytes have activated adhesion receptors. PSGL-1 at the uropod and Mac-1 at the leading edge are putatively involved in interactions with platelets on polarized cells. Interactions between the leading edge and circulating RBCs are also mediated by the leading edge. This capture is mediated by ligands on platelets and RBCs that have not yet been identified. By promoting ROS release, engaged integrins deliver outside-in signals to adherent neutrophils. (2) Platelet interactions. In response to stimulated or damaged endothelium, resting or activated platelets can firmly arrest. Resting platelets express GPIbα and PSGL-1 which mediate the tethering on vWf and P-selectin on the endothelium, respectively. GPIIb/IIIa promotes aggregation of activated platelets or adhesion to vessel walls during inflammation. It has been shown that adherent platelets can enhance pathologicalconditions through P-selectin-mediated recruitment of leukocytes. Binding to leukocytes in the circulation is mediated by P-selectin and GPIbα, uncharacterized receptor/ligand pairs.*

Source: Maria Nacher, & Andes Hidalgo, (2011). Fire within the vessels: interactions between blood cells and inflammatory vascular injury. Front. Biosci. (Schol Ed), 3(3), 1089–1100.

2.16. IMPLICATIONS IN CARDIOVASCULAR DISEASES

A significant correlation has nonetheless been found between blood leukocyte counts and long-term vascular disease caused by atherosclerosis, diabetes, or SCD. Neutrophils showed the strongest correlations among other subsets of leukocytes.

It is likely that neutrophil activation in affected vascular beds exacerbates endothelial injury and preserves inflammation even though other myeloid leukocyte types and cells of the adaptive immune system are involved in the initiation of these inflammatory disorders (Woollard&Geissmann, 2010). Cardiovascular disease risk has indeed been linked to increased levels of leukocyte-platelet aggregates. According to these data, hyperlipidemic patients have elevated levels of neutrophil-derived myeloperoxidase and primed neutrophils (Mazor et al., 2008).

In the ApoE null mouse model, Klaus Ley and colleagues elegantly uncovered one plausible mechanism by which leukocyte-platelet interactions promote atherosclerosis (Huo et al., 2003). By delivering proinflammatory chemokines to the inflamed or atherosclerotic endothelium, activated platelets bind circulating monocytes in a P-selectin-dependent manner. Consequently, ApoE null mice's aortas showed an increase in lesion size (Huo et al., 2003). In recent studies, neutrophils have also been found to accumulate in significant numbers within and outside atherosclerotic lesions (Rotzius et al., 2010). Christian Weber's group reported that diet-induced atherosclerosis in ApoE-null mice was aggravated by the forced egress of bone marrow neutrophils, but was attenuated by their depletion (Zernecke et al., 2008).

It remains largely unknown whether selectin-dependent, local activation of neutrophils that engage circulating platelets enhances endothelial injury and contributes to plaque formation. This contention is further supported by a number of observations, including the expression of E-selectin in endothelial beds associated with human atherosclerotic plaques, its contribution to plaque formation alone or in combination with P-selectin (Collins et al., 2000), and the significant role of platelet-derived P-selectin (Burger & Wagner, 2003) and beta2-integrins on leukocytes in promoting long-term vascular injury, including atherosclerotic plaque formation and growth.

By using improved imaging technologies and relevant animal models, future research may allow us to unravel how interactions among blood cells lead to acute and chronic cardiovascular disease, and could lead to the rational design of therapies to combat this group of complications that so overwhelmingly threaten our lives (Nacher& Hidalgo, 2011).

2.17. CONCLUSION

This article summarizes our current knowledge of RBC interactions with other cells, ECs, and platelets in physiological and disease-related conditions. There have been both direct and indirect interactions between RBCs, ECs, platelets, WBCs, macrophages, and other RBCs discussed. There are indirect interactions that can occur via plasma ligands, proteins, and molecules released by these cells. This chapter describes other indirect interactions as mechanical: these involve RBCs in contact with other cells under physiological flow conditions and their dynamic and rheological distribution. It illustrates the complexity of global interactions in which mature RBCs participate, and more importantly, it emphasizes the importance of addressing pathological conditions. Interactions with the endothelium are largely influenced by depletion interaction, while interactions with platelets and WBCs are largely influenced by RBC margination. In addition, we discussed RBC's critical interactions with pathogens.

In most cases, research on RBC interactions with other cells is conducted under artificial conditions that may differ from physiological ones. Due to the fact that cultured ECs have a different structure and amount of glycocalyx than *in vivo* vascular ECs, the interaction with cultured ECs may differ from that with *in vivo* vascular ECs, for example (Weinbaum et al., 2007). Additionally, in experimental setups, cells present *in vivo* are often excluded to avoid influencing measurements. The influence of the three-dimensional vessel wall is not taken into account in *in vitro* studies because they are often conducted with flat cultured monolayers of ECs. SCD patients' blood or rodent blood is used in the majority of studies examining RBC adhesiveness. Their membranes contain PS and other receptors that enable them to adhere to other cells. In contrast, sickle cells are not known to adhere in the same way as other activated RBCs. Further, RBC adhesion is likely mediated by a number of receptors and interactions. Furthermore, it does not necessarily follow that high-affinity adhesion mechanisms are the most important factor *in vivo* adhesion. Granulocytes, which are larger than RBCs, can cause intermittent microvascular blood flow and slow the flow of RBCs. Adhesion may also be influenced by low-affinity mechanisms in such conditions.

RBCs are recurrently observed to become more adhesive and less deformable in a variety of pathological conditions, e.g., SCD, T2DM, G6PD deficiency, thalassemia, chronic diseases, and malaria, which facilitates cell-to-cell interactions. There is a consequent impairment in oxygen and nutrient delivery to organs and tissues as a result of the development of microvascular obstructions. Finally, this chapter expands our understanding of RBC function and role beyond the characteristics of the RBC itself. We can improve our understanding of RBC homeostasis by studying the interactions between RBCs and other cell types, proteins, pathogens, and cellular mediators. Blood cells interact with endothelium, but the system is more complex and includes plasma components, blood cells, and vessel wall reactions. In addition to treating dyslipidemia, hypercoagulability, and other metabolic disorders, the therapeutic approaches target plasmatic abnormalities as well. Furthermore, infections and immunological dysfunctions contribute to vascular abnormalities and can be prevented and treated. By using humanized specific monoclonal antibodies, peptides, and gene transcription modulators, we would be able to design future treatments with better outcomes.

"That a universe exists within every human being. That to the blood cells and organs in your body, you are god. That this universe is only one individual among infinite others."

— *Peter Tieryas, United States of Japan*

REFERENCES

1. Anderson, H. L., Brodsky, I., &Mangalmurti, N. (2018). The evolving erythrocyte: Red blood cells as modulators of innate immunity. *Journal of Immunology, 201*, 1343–1351.
2. Andre, P., Denis, C. V., Ware, J., Saffaripour, S., Hynes, R. O., Ruggeri, Z. M., & Wagner, D. D. (2000). Platelets adhere to and translocate on von Willebrand factor presented by endothelium in stimulated veins. *Blood, 96*(10), 3322–3328.
3. Andrews, R. K., & Berndt, M. C. (2013). Bernard-Soulier syndrome: An update. *Seminars in Thrombosis and Hemostasis, 39*, 656–662.
4. Auffray, C., Fogg, D., Garfa, M., Elain, G., Join-Lambert, O., Kayal, S., Sarnacki, S., Cumano, A., Lauvau, G., &Geissmann, F. (2007). Monitoring of blood vessels and tissues by a population of monocytes with patrolling behavior. *Science, 317*(5838), 666–670.
5. Bächer, C., Kihm, A., Schrack, L., Kaestner, L., Laschke, M., Wagner, C., et al. (2018). Antimargination of microparticles and platelets in the vicinity of branching vessels. *Biophysical Journal, 115*, 411–425.
6. Badior, K., & Casey, J. (2018). Molecular mechanism for the red blood cell senescence clock. *IUBMB Life, 70*, 32–40.
7. Barbe, L. L., Boval, B. M., Wautier, M. P., &Wautier, J. L. (2001). An *in vitro* system for testing leucocyte and leukemic cell line adhesion to synthetic fibers. *British Journal of Haematology, 115*, 664–671.
8. Barreiro, O., de la Fuente, H., Mittelbrunn, M., & Sanchez-Madrid, F. (2007). Functional insights on the polarized redistribution of leukocyte integrins and their ligands during leukocyte migration and immune interactions. *Immunological Reviews, 218*, 147–164.
9. Baum, J., Ward, R., & Conway, D. (2002). Natural selection on the erythrocyte surface. *Molecular Biology and Evolution, 19*, 223–229.
10. Bazzoni, G., &Dejana, E. (2004). Endothelial cell-to-cell junctions: Molecular organization and role in vascular homeostasis. *Physiological Reviews, 84*(3), 869–901.
11. Beck, Z., Brown, B., Wieczorek, L., Peachman, K., Matyas, G., Polonis, V., et al. (2009). Human erythrocytes selectively bind and enrich infectious HIV-1 virions. *PLoS One, 4*(8), e8297.
12. Bishop, J. J., Nance, P., Popel, A. S., Intaglietta, M., & Johnson, P. C. (2001). Effect of erythrocyte aggregation on velocity profiles in venules. *American Journal of Physiology-Heart and Circulatory Physiology, 280*, H222–H236.

13. Brinkmann, V., Reichard, U., Goosmann, C., Fauler, B., Uhlemann, Y., Weiss, D. S., Weinrauch, Y., &Zychlinsky, A. (2004). Neutrophil extracellular traps kill bacteria. *Science, 303*(5663), 1532–1535.
14. Brown, E. B., Campbell, R. B., Tsuzuki, Y., Xu, L., Carmeliet, P., Fukumura, D., & Jain, R. K. (2001). *In vivo* measurement of gene expression, angiogenesis and physiological function in tumors using multiphoton laser scanning microscopy. *Nature Medicine, 7*, 864–868.
15. Brownlee, M. (2001). Biochemistry and molecular cell biology of diabetic complications. *Nature, 414*, 813–820.
16. Burger, P. C., & Wagner, D. D. (2003). Platelet P-selectin facilitates atherosclerotic lesion development. *Blood, 101*(7), 2661–2666.
17. Carboni, E. J. (2017). *The Margination and Transport of Particles in Blood Flow* (Doctoral dissertation). University of Connecticut, Storrs, CT.
18. Chang, H., Yazdani, A., Li, X., Douglas, K., Mantzoros, C., &Karniadakis, G. (2018). Quantifying platelet margination in diabetic blood flow. *Biophysical Journal, 115*, 1371–1382.
19. Chapman, G. B., &Cokelet, G. R. (1998). Flow resistance and drag forces due to multiple adherent leukocytes in post-capillary vessels. *Biophysical Journal, 74*, 3292–3301.
20. Chen, B. P., Li, Y. S., Zhao, Y., Chen, K. D., Li, S., Lao, J., Yuan, S., Shyy, J. Y., &Chien, S. (2001). DNA microarray analysis of gene expression in endothelial cells in response to 24-h shear stress. *Physiological Genomics, 7*, 55–63.
21. Chiang, E. Y., Hidalgo, A., Chang, J., &Frenette, P. S. (2007). Imaging receptor microdomains on leukocyte subsets in live mice. *Nature Methods, 4*(3), 219–222.
22. Clark, S. R., Ma, A. C., Tavener, S. A., McDonald, B., Goodarzi, Z., Kelly, M. M., Patel, K. D., Chakrabarti, S., McAvoy, E., Sinclair, G. D., Keys, E. M., Allen-Vercoe, E., Devinney, R., Doig, C. J., Green, F. H., &Kubes, P. (2007). Platelet TLR4 activates neutrophil extracellular traps to ensnare bacteria in septic blood. *Nature Medicine, 13*(4), 463–469. https://doi.org/10.1038/nm1565.
23. Cokelet, G. R., & Goldsmith, H. L. (1991). Decreased hydrodynamic resistance in the two-phase flow of blood through small vertical tubes at low flow rates. *Circulation Research, 68*, 1–17. https://doi.org/10.1161/01.RES.68.1.1.
24. Collins, R. G., Velji, R., Guevara, N. V., Hicks, M. J., Chan, L., & Beaudet, A. L. (2000). P-Selectin or intercellular adhesion molecule (ICAM)-1

deficiency substantially protects against atherosclerosis in apolipoprotein E-deficient mice. *Journal of Experimental Medicine, 191*(1), 189–194. https://doi.org/10.1084/jem.191.1.189.

25. Das, B., Johnson, P., &Popel, A. S. (2000). Computational fluid dynamic studies of leukocyte adhesion effects on non-Newtonian blood flow through microvessels. *Biorheology, 37*, 239–258.
26. Dupin, M. M., Halliday, I., Care, C. M., Alboul, L., & Munn, L. L. (2007). Modeling the flow of dense suspensions of deformable particles in three dimensions. *Physical Review E, 75*, 066707. https://doi.org/10.1103/PhysRevE.75.066707.
27. El Nemer, W., Wautier, M. P., Rahuel, C., Gane, P., Hermand, P., Galacteros, F., Wautier, J. L., Cartron, J. P., Colin, Y., & Le Van Kim, C. (2007). Endothelial Lu/BCAM glycoproteins are novel ligands for red blood cell alpha4beta1 integrin: Role in adhesion of sickle red blood cells to endothelial cells. *Blood, 109*, 3544–3551.
28. Etzioni, A. (2009). Genetic etiologies of leukocyte adhesion defects. *Current Opinion in Immunology, 21*(5), 481–486.
29. Fedosov, D. A., &Gompper, G. (2014). White blood cells margination in microcirculation. *RSC Advances, 10*, 2961–2970.
30. Filippi, M. D., Szczur, K., Harris, C. E., &Berclaz, P. Y. (2007). Rho GTPase Rac1 is critical for neutrophil migration into the lung. *Blood, 109*(3), 1257–1264.
31. Finnegan, E. M., Turhan, A., Golan, D., &Barabino, G. (2007). Adherent leukocytes capture sickle erythrocytes in an *in vitro* flow model of Vaso occlusion. *American Journal of Hematology, 82*, 266–275.
32. Fitzgibbon, S., Spann, A., Qi, Q., &Shaqfeh, E. (2015). *In vitro* measurement of particle margination in the microchannel flow: Effect of varying hematocrit. *Biophysical Journal, 108*, 2601–2608.
33. Flegel, W. A. (2015). Pathogenesis and mechanisms of antibody-mediated hemolysis. *Transfusion, 55*, 47–58.
34. Frenette, P. S. (2004). Sickle cell Vaso occlusion: Heterotypic, multicellular aggregations driven by leukocyte adhesion. *Microcirculation, 11*(2), 167–177.
35. Frenette, P. S., &Atweh, G. F. (2007). Sickle cell disease: Old discoveries, new concepts, and future promise. *Journal of Clinical Investigation, 117*(4), 850–858.
36. Furchgott, R. F., &Zawadzki, J. V. (1980). The obligatory role of endothelial cells in the relaxation of arterial smooth muscle by acetylcholine. *Nature, 288*, 373–376. https://doi.org/10.1038/288373a0.

37. Gimbrone, M. A., Jr., & Garcia-Cardena, G. (2016). Endothelial cell dysfunction and the pathobiology of atherosclerosis. *Circulation Research, 118*, 620–636. https://doi.org/10.1161/CIRCRESAHA.115.306301.

38. Goel, M. S., & Diamond, S. L. (2002). Adhesion of normal erythrocytes at depressed venous shear rates to activated neutrophils, activated platelets, and fibrin polymerized from plasma. *Blood, 100*(10), 3797–3803.

39. Goldsmith, H. L., Cokelet, G. R., &Gaehtgens, P. (1989). Robin Fahraeus: Evolution of his concepts in cardiovascular physiology. *American Journal of Physiology, 257*, H1005–H1015.

40. Grossin, N., Wautier, M. P., Picot, J., Stern, D. M., &Wautier, J. L. (2009). Differential effect of plasma or erythrocyte AGE-ligands of RAGE on expression of transcripts for receptor isoforms. *Diabetes & Metabolism, 35*, 410–417.

41. Hebbel, R. P., Boogaerts, M. A., Eaton, J. W., & Steinberg, M. H. (1980). Erythrocyte adherence to endothelium in sickle-cell anemia. A possible determinant of disease severity. *New England Journal of Medicine, 302*, 992–995.

42. Heit, B., Robbins, S. M., Downey, C. M., Guan, Z., Colarusso, P., Miller, B. J., Jirik, F. R., &Kubes, P. (2008). PTEN functions to 'prioritize' chemotactic cues and prevent 'distraction' in migrating neutrophils. *Nature Immunology, 9*(7), 743–752.

43. Hidalgo, A., Chang, J., Jang, J. E., Peired, A. J., Chiang, E. Y., &Frenette, P. S. (2009). Heterotypic interactions enabled by polarized neutrophil microdomains mediate thromboinflammatory injury. *Nature Medicine, 15*(4), 384–391. https://doi.org/10.1038/nm.1939.

44. Hidalgo, A., Peired, A. J., Wild, M. K., Vestweber, D., &Frenette, P. S. (2007). Complete identification of E-selectin ligands on neutrophils reveals distinct functions of PSGL-1, ESL-1, and CD44. *Immunity, 26*(4), 477–489. https://doi.org/10.1016/j.immuni.2007.03.011.

45. Hudson, B. I., Carter, A. M., Harja, E., Kalea, A. Z., Arriero, M., Yang, H., Grant, P. J., & Schmidt, A. M. (2008). Identification, classification, and expression of RAGE gene splice variants. *The FASEB Journal, 22*, 1572–1580.

46. Huo, Y., Schober, A., Forlow, S. B., Smith, D. F., Hyman, M. C., Jung, S., Littman, D. R., Weber, C., & Ley, K. (2003). Circulating activated platelets exacerbate atherosclerosis in mice deficient in apolipoprotein E. *Nature Medicine, 9*(1), 61–67. https://doi.org/10.1038/nm810.

47. Hynes, R. O. (1987). Integrins: A family of cell surface receptors. *Cell, 48*, 549–554.

48. Jackson, S. P. (2007). The growing complexity of platelet aggregation. *Blood, 109*(12), 5087–5095.

49. James, C., Ugo, V., Le Couedic, J. P., Staerk, J., Delhommeau, F., Lacout, C., Garcon, L., Raslova, H., Berger, R., Bennaceur-Griscelli, A., et al. (2005). A unique clonal *JAK2* mutation leading to constitutive signaling causes polycythemia vera. *Nature, 434*, 1144–1148. https://doi.org/10.1038/nature03546.

50. Janowska-Wieczorek, A., Majka, M., Kijowski, J., Baj-Krzyworzeka, M., Reca, R., Turner, A. R., Ratajczak, J., Emerson, S. G., Kowalska, M. A., &Ratajczak, M. Z. (2001). Platelet-derived microparticles bind to hematopoietic stem/progenitor cells and enhance their engraftment. *Blood, 98*(10), 3143–3149. https://doi.org/10.1182/blood.V98.10.3143.

51. Jones, D. P., True, H. D., & Patel, J. (2017). Leukocyte trafficking in cardiovascular disease: Insights from experimental models. *Mediators of Inflammation, 2017*, 9746169. https://doi.org/10.1155/2017/9746169.

52. Kaliyaperumal, R., Deng, X., Meiselman, H., Song, H., Dalan, R., Leow, M., et al. (2019). Depletion interaction forces contribute to erythrocyte-endothelial adhesion in diabetes. *Biochemical and Biophysical Research Communications, 516*, 144–148.

53. Kieffer, N., Wautier, J. L., Coulombel, L., Titeux, M., Wautier, M. P., Vainchenker, W., Ruan, C., & Breton-Gorius, J. (1988). Uncoupling in the expression of platelet GP IIb/IIIa in human endothelial cells and K562 cells: Absence of immunologic crossreactivity between platelet GP IIb and the vitronectin receptor alpha chain. *Blood, 72*, 1209–1215.

54. Kiladjian, J. J., Cervantes, F., Leebeek, F. W., Marzac, C., Cassinat, B., Chevret, S., Cazals-Hatem, D., Plessier, A., Garcia-Pagan, J. C., Darwish Murad, S., et al. (2008). The impact of JAK2 and MPL mutations on diagnosis and prognosis of splanchnic vein thrombosis: A report on 241 cases. *Blood, 111*, 4922–4929.

55. Kinashi, T. (2005). Intracellular signaling controlling integrin activation in lymphocytes. *Nature Reviews Immunology, 5*(7), 546–559. https://doi.org/10.1038/nri1646.

56. Klei, T., Meinderts, S., van den Berg, T., & van Bruggen, R. (2017). From the cradle to the grave: The role of macrophages in erythropoiesis and erythrophagocytosis. *Frontiers in Immunology, 8*, 73. https://doi.org/10.3389/fimmu.2017.00073.

57. Klein, R., Klein, B. E., Moss, S. E., &Meuer, S. M. (2000). The epidemiology of retinal vein occlusion: The Beaver Dam Eye Study. *Transactions of the American Ophthalmological Society, 98*, 133–141.

58. Kroll, M. H., Hellums, J. D., McIntire, L. V., Schafer, A. I., &Moake, J. L. (1996). Platelets and shear stress. *Blood, 88*, 1525–1541.

59. Langer, H. F., Bigalke, B., Seizer, P., Stellos, K., Fateh-Moghadam, S., &Gawaz, M. (2010). Interaction of platelets and inflammatory endothelium in the development and progression of coronary artery disease. *Seminars in Thrombosis and Hemostasis, 36*(2), 131–138.

60. Lee, J. H., Katakai, T., Hara, T., Gonda, H., Sugai, M., & Shimizu, A. (2004). Roles of p-ERM and Rho-ROCK signaling in lymphocyte polarity and uropod formation. *Journal of Cell Biology, 167*(2), 327–337.

61. Lee, K., Gane, P., Roudot-Thoraval, F., Godeau, B., Bachir, D., Bernaudin, F., Cartron, J. P., Galacteros, F., &Bierling, P. (2001). The nonexpression of CD36 on reticulocytes and mature red blood cells does not modify the clinical course of patients with sickle cell anemia. *Blood, 98*, 966–971. https://doi.org/10.1182/blood.V98.4.966

62. Li, N., Ji, Q., &Hjemdahl, P. (2006). Platelet-lymphocyte conjugation differs between lymphocyte subpopulations. *Journal of Thrombosis and Haemostasis, 4*(4), 874–881.

63. Libby, P. (2013). Mechanisms of acute coronary syndromes and their implications for therapy. *New England Journal of Medicine, 368*, 2004–2013. https://doi.org/10.1056/NEJMra1216063.

64. Liote, F., Boval-Boizard, B., Weill, D., Kuntz, D., &Wautier, J. L. (1996). Blood monocyte activation in rheumatoid arthritis: Increased monocyte adhesiveness, integrin expression, and cytokine release. *Clinical and Experimental Immunology, 106*, 13–19.

65. Liu, M., & Yang, J. (2009). Electrokinetic effect of the endothelial glycocalyx layer on two-phase blood flow in small blood vessels. *Microvascular Research, 78*, 14–19.

66. Looney, M. R., Gilliss, B. M., &Matthay, M. A. (2010). Pathophysiology of transfusion-related acute lung injury. *Current Opinion in Hematology, 17*(5), 418–423.

67. Looney, M. R., Nguyen, J. X., Hu, Y., Van Ziffle, J. A., Lowell, C. A., &Matthay, M. A. (2009). Platelet depletion and aspirin treatment protect mice in a two-event model of transfusion-related acute lung injury. *Journal of Clinical Investigation, 119*(11), 3450–3461. https://doi.org/10.1172/JCI38432.

68. Massberg, S., Konrad, I., Bultmann, A., Schulz, C., Munch, G., Peluso, M., Lorenz, M., Schneider, S., Besta, F., Muller, I., Hu, B., Langer, H., Kremmer, E., Rudelius, M., Heinzmann, U., Ungerer, M., &Gawaz, M. (2004). Soluble glycoprotein VI dimer inhibits platelet adhesion and

aggregation to the injured vessel wall *in vivo*. *The FASEB Journal, 18*(2), 397–399.

69. Maugeri, N., Rovere-Querini, P., Evangelista, V., Covino, C., Capobianco, A., Bertilaccio, M. T., Piccoli, A., Totani, L., Cianflone, D., Maseri, A., & Manfredi, A. A. (2009). Neutrophils phagocytose activated platelets *in vivo*: A phosphatidylserine, P-selectin, and β2 integrin-dependent cell clearance program. *Blood, 113*(21), 5254–5265.
70. Mazor, R., Shurtz-Swirski, R., Farah, R., Kristal, B., Shapiro, G., Dorlechter, F., Cohen-Mazor, M., Meilin, E., & Sela, S. (2008). Primed polymorphonuclear leukocytes constitute a possible link between inflammation and oxidative stress in hyperlipidemic patients. *Atherosclerosis, 197*(2), 937–943.
71. McDonald, M. J., Shapiro, R., Bleichman, M., Solway, J., & Bunn, H. F. (1978). Glycosylated minor components of human adult hemoglobin. Purification, identification, and partial structural analysis. *Journal of Biological Chemistry, 253*, 2327–2332.
72. McEver, R. P., & Zhu, C. (2010). Rolling cell adhesion. *Annual Review of Cell and Developmental Biology, 26*, 363–396.
73. McMorran, B. J., Marshall, V. M., de Graaf, C., Drysdale, K. E., Shabbar, M., Smyth, G. K., Corbin, J. E., Alexander, W. S., & Foote, S. J. (2009). Platelets kill intraerythrocytic malarial parasites and mediate survival to infection. *Science, 323*(5915), 797–800. https://doi.org/10.1126/science.1166296.
74. Medzhitov, R. (2008). Origin and physiological roles of inflammation. *Nature, 454*(7203), 428–435. https://doi.org/10.1038/nature07201.
75. Migliorini, C., Qian, Y., Chen, H., Brown, E., Jain, R., & Munn, L. (2002). Red blood cells augment leukocyte rolling in a virtual blood vessel. *Biophysical Journal, 83*(4), 1834–1841.
76. Minasyan, H. (2016). Mechanisms and pathways for the clearance of bacteria from blood circulation in health and disease. *Pathophysiology, 23*, 61–66.
77. Minasyan, H. (2018). Phagocytosis and oxycytosis: Two arms of human innate immunity. *Immunologic Research, 66*, 271–280.
78. Morera, D., &MacKenzie, S. A. (2011). Is there a direct role for erythrocytes in the immune response? *Veterinary Research, 42*, 89. https://doi.org/10.1186/1297-9716-42-89.
79. Moser, M., Bauer, M., Schmid, S., Ruppert, R., Schmidt, S., Sixt, M., Wang, H. V., Sperandio, M., &Fassler, R. (2009). Kindlin-3 is required

for β2 integrin-mediated leukocyte adhesion to endothelial cells. *Nature Medicine, 15*(3), 300–305. https://doi.org/10.1038/nm.1921.

80. Motto, D. G., Chauhan, A. K., Zhu, G., Homeister, J., Lamb, C. B., Desch, K. C., Zhang, W., Tsai, H. M., Wagner, D. D., & Ginsburg, D. (2005). Shigatoxin triggers thrombotic thrombocytopenic purpura in genetically susceptible ADAMTS13-deficient mice. *Journal of Clinical Investigation, 115*(10), 2752–2761.
81. Munn, L. L., &Dupin, M. M. (2008). Blood cell interactions and segregation in flow. *Annals of Biomedical Engineering, 36*(4), 534–544.
82. Munn, L., Mulivor, A. W., Dupin, M., Shevkoplyas, S. S., & Sun, C. (2006). Tumor vessel abnormalities affect blood cell dynamics and flow distribution. *Journal of Biomechanics, 39*, S396.
83. Nacher, M., & Hidalgo, A. (2011). Fire within the vessels: Interactions between blood cells and inflammatory vascular injury. *Frontiers in Bioscience (Scholar Edition), 3*(3), 1089–1100.
84. Nentwich, M. M., &Ulbig, M. W. (2015). Diabetic retinopathy—Ocular complications of diabetes mellitus. *World Journal of Diabetes, 6*, 489–499. https://doi.org/10.4239/wjd.v6.i3.489.
85. Nesbitt, W. S., Kulkarni, S., Giuliano, S., Goncalves, I., Dopheide, S. M., Yap, C. L., Harper, I. S., Salem, H. H., & Jackson, S. P. (2002). Distinct glycoprotein Ib/V/IX and integrin αIIbβ3-dependent calcium signals cooperatively regulate platelet adhesion under flow. *Journal of Biological Chemistry, 277*(4), 2965–2972.
86. Neu, B., & Meiselman, H. (2006). Depletion interactions in polymer solutions promote red blood cell adhesion to albumin-coated surfaces. *Biochimica et Biophysica Acta (BBA) - General Subjects, 1760*, 1772–1779.
87. Neu, B., Sowemimo-Coker, S. O., & Meiselman, H. J. (2003). Cell-cell affinity of senescent human erythrocytes. *Biophysical Journal, 85*, 75–84.
88. Nishikimi, A., Fukuhara, H., Su, W., Hongu, T., Takasuga, S., Mihara, H., Cao, Q., Sanematsu, F., Kanai, M., Hasegawa, H., Tanaka, Y., Shibasaki, M., Kanaho, Y., Sasaki, T., Frohman, M. A., & Fukui, Y. (2009). Sequential regulation of DOCK2 dynamics by two phospholipids during neutrophil chemotaxis. *Science, 324*(5925), 384–387.
89. Nurden, A. T., Pillois, X., & Wilcox, D. A. (2013). Glanzmann thrombasthenia: State of the art and future directions. *Seminars in Thrombosis and Hemostasis, 39*, 642–655.
90. Oberleithner, H. (2013). Vascular endothelium leaves fingerprints on the

surface of erythrocytes. *Molecular and Cellular Mechanisms of Disease, 465*, 1451–1458.

91. Oberleithner, H., Peters, W., Kusche-Vihrog, K., Korte, S., Schillers, H., Kliehe, K., et al. (2011). Salt overload damages the glycocalyx sodium barrier of vascular endothelium. *PflugersArchiv: European Journal of Physiology, 462*, 519–528.
92. Palmblad, J., &Papadaki, H. A. (2008). Chronic idiopathic neutropenia's and severe congenital neutropenia. *Current Opinion in Hematology, 15*(1), 8–14.
93. Parroche, P., Lauw, F., Goutagny, E., Monks, B., Visentin, A., Halmen, K., et al. (2007). Malaria hemozoin is immunologically inert but radically enhances innate responses by presenting malaria DNA to Toll-like receptor 9. *Proceedings of the National Academy of Sciences of the United States of America, 104*, 1919–1924. https://doi.org/10.1073/pnas.0608745104.
94. Patel, D., Vaananen, H., Jirouskova, M., Hoffmann, T., Bodian, C., &Coller, B. S. (2003). Dynamics of GPIIb/IIIa-mediated platelet-platelet interactions in platelet adhesion/thrombus formation on collagen *in vitro* as revealed by video microscopy. *Blood, 101*(3), 929–936.
95. Pearson, M. J., &Lipowsky, H. H. (2000). Influence of erythrocyte aggregation on leukocyte margination in postcapillary venules of rat mesentery. *American Journal of Physiology-Heart and Circulatory Physiology, 279*, H1460–H1471. https://doi.org/10.1152/ajpheart.2000.279.4.H1460.
96. Peyvandi, F., Garagiola, I., &Baronciani, L. (2011). Role of von Willebrand factor in the haemostasis. *Blood Transfusion, 9*, s3–s8.
97. Prevost, G., Fajardy, I., Besmond, C., Balkau, B., Tichet, J., Fontaine, P., Danze, P. M., Marre, M., et al. (2005). Polymorphisms of the receptor of advanced glycation endproducts (RAGE) and the development of nephropathy in type 1 diabetic patients. *Diabetes & Metabolism, 31*, 35–39.
98. Rainger, G. E., Chimen, M., Harrison, M. J., Yates, C. M., Harrison, P., Watson, S. P., Lordkipanidze, M., & Nash, G. B. (2015). The role of platelets in the recruitment of leukocytes during vascular disease. *Platelets, 26*, 507–520.
99. Ren, Q., Ye, S., & Whiteheart, S. W. (2008). The platelet release reaction: Just when you thought platelet secretion was simple. *Current Opinion in Hematology, 15*(5), 537–-541.
100. Rotzius, P., Thams, S., Soehnlein, O., Kenne, E., Tseng, C. N., Bjorkstrom, N. K., Malmberg, K. J., Lindbom, L., & Eriksson, E. E. (2010). Distinct

infiltration of neutrophils in lesion shoulders in ApoE-/- mice. *American Journal of Pathology, 177*(1), 493–500.

101. Schleicher, E. D., Wagner, E., &Nerlich, A. G. (1997). Increased accumulation of the glycoxidation product N(epsilon)-(carboxymethyl) lysine in human tissues in diabetes and aging. *Journal of Clinical Investigation, 99*, 457–468.
102. Schmid-Schoenbein, G. W., Usami, S., Skalak, R., &Chien, S. (1980). The interaction of leukocytes and erythrocytes in capillary and postcapillary vessels. *Microvascular Research, 19*, 45–70.
103. Semple, J. W., & Freedman, J. (2010). Platelets and innate immunity. *Cellular and Molecular Life Sciences, 67*(4), 499–511.
104. Setty, Y., Kulkarni, S., & Stuart, M. J. (2002). Role of erythrocyte phosphatidylserine in sickle red cell-endothelial adhesion. *Blood, 99*, 1564–1571. https://doi.org/10.1182/blood.V99.5.1564.
105. Shipanova, I. N., Glomb, M. A., & Nagaraj, R. H. (1997). Protein modification by methylglyoxal: Chemical nature and synthetic mechanism of a major fluorescent adduct. *Archives of Biochemistry and Biophysics, 344*, 29–36.
106. Simon, S. I., & Goldsmith, H. L. (2002). Leukocyte adhesion dynamics in shear flow. *Annals of Biomedical Engineering, 30*, 315–332.
107. Southcott, M. J., Tanner, M. J., &Anstee, D. J. (1999). The expression of human blood group antigens during erythropoiesis in a cell culture system. *Blood, 93*, 4425–4435. https://doi.org/10.1182/blood.V93.12.4425.
108. Sperandio, M., Smith, M. L., Forlow, S. B., Olson, T. S., Xia, L., McEver, R. P., & Ley, K. (2003). P-selectin glycoprotein ligand-1 mediates L-selectin-dependent leukocyte rolling in venules. *Journal of Experimental Medicine, 197*(10), 1355–1363.
109. Stellos, K., Gnerlich, S., Kraemer, B., Lindemann, S., &Gawaz, M. (2008). Platelet interaction with progenitor cells: Vascular regeneration or inquiry? *Pharmacological Reports, 60*(1), 101–108. PMID: 18320245
110. Stone, M., Bakkour, S., Lee, H. T., Lanteri, M., Simmons, G., Brambilla, D., et al. (2017). Zika RNA persistence in blood and body fluids and clinical outcomes in infected blood donors. *Transfusion, 57*(4A), 4A. PMID: 28328055.
111. Sun, C. H., & Munn, L. L. (2005). Particulate nature of blood determines macroscopic rheology: A 2-D lattice Boltzmann analysis. *Biophysical Journal, 88*, 1635–1645.
112. Sun, C. H., & Munn, L. L. (2006). Influence of erythrocyte aggregation on

leukocyte margination in postcapillary expansions: A lattice-Boltzmann analysis. *Physica A: Statistical Mechanics and its Applications, 362*, 191–196.

113. Turhan, A., Weiss, L. A., Mohandas, N., Coller, B. S., &Frenette, P. S. (2002). Primary role for adherent leukocytes in sickle cell vascular occlusion: A new paradigm. *Proceedings of the National Academy of Sciences of the United States of America, 99*(5), 3047–3051.
114. Vahidkhah, K., Diamond, S. L., &Bagchi, P. (2013). Hydrodynamic interaction between a platelet and an erythrocyte: Effect of erythrocyte deformability, dynamics, and wall proximity. *Journal of Biomechanical Engineering, 135*(5), 051002.
115. Vahidkhah, K., Diamond, S. L., &Bagchi, P. (2014). Platelet dynamics in three-dimensional simulation of whole blood. *Biophysical Journal, 106*, 2529–2540.
116. Vannucchi, A. M., Antonioli, E., Guglielmelli, P., Longo, G., Pancrazzi, A., Ponziani, V., et al. (2007). Prospective identification of high-risk polycythemia vera patients based on JAK2(V617F) allele burden. *Leukemia, 21*, 1952–1959.
117. Varma, D. D., Cugati, S., Lee, A. W., & Chen, C. S. (2013). A review of central retinal artery occlusion: Clinical presentation and management. *Eye, 27*, 688–697.
118. Wagner, D. D., &Frenette, P. S. (2008). The vessel wall and its interactions. *Blood, 111*(11), 5271–5281. https://doi.org/10.1182/blood-2008-01-078204.
119. Walton, L., Lehmann, M., Wkorczewski, T., Holle, L. A., Beckman, J. D., Cribb, J. A., et al. (2017). Elevated hematocrit enhances platelet accumulation following vascular injury. *Blood, 129*, 2537–2546.
120. Wang, H. B., Wang, J. T., Zhang, L., Geng, Z. H., Xu, W. L., Xu, T., et al. (2007). P-selectin primes leukocyte integrin activation during inflammation. *Nature Immunology, 8*(8), 882–892.
121. Wang, Y., Sakuma, M., Chen, Z., Ustinov, V., Shi, C., Croce, K., et al. (2005). Leukocyte engagement of platelet glycoprotein Ibalpha via the integrin Mac-1 is critical for the biological response to vascular injury. *Circulation, 112*(19), 2993–3000.
122. Wautier, J. L., &Guillausseau, P. J. (1998). Diabetes, advanced glycation endproducts and vascular disease. *Vascular Medicine, 3*(2), 131–137. https://doi.org/10.1177/1358836X9800300207.
123. Wautier, J. L., & Schmidt, A. M. (2004). Protein glycation: A firm link

to endothelial cell dysfunction. *Circulation Research, 95*(3), 233–238. https://doi.org/10.1161/01.RES.0000137876.28454.64.

124. Wautier, J. L., Galacteros, F., Wautier, M. P., Pintigny, D., Beuzard, Y., Rosa, J., & Caen, J. P. (1985). Clinical manifestations and erythrocyte adhesion to endothelium in sickle cell syndrome. *American Journal of Hematology, 19*(2), 121–130.

125. Wautier, J.-L., &Wautier, M.-P. (2020). Cellular and molecular aspects of blood cell–endothelium interactions in vascular disorders. *International Journal of Molecular Sciences, 21*(15), 5315. https://doi.org/10.3390/ijms21155315.

126. Wautier, M. P., El Nemer, W., Gane, P., Rain, J. D., Cartron, J. P., Colin, Y., Le Van Kim, C., &Wautier, J. L. (2007). Increased adhesion to endothelial cells of erythrocytes from patients with polycythemia vera is mediated by laminin alpha5 chain and Lu/BCAM. *Blood, 110*(3), 894–901.

127. Wautier, M. P., Heron, E., Picot, J., Colin, Y., Hermine, O., &Wautier, J. L. (2011). Red blood cell phosphatidylserine exposure is responsible for increased erythrocyte adhesion to endothelium in central retinal vein occlusion. *Journal of Thrombosis and Haemostasis, 9*(5), 1049–1055.

128. Weinbaum, S., Tarbell, J. M., & Damiano, E. R. (2007). The structure and function of the endothelial glycocalyx layer. *Annual Review of Biomedical Engineering, 9*, 121–167. https://doi.org/10.1146/annurev.bioeng.9.060906.151959.

129. Woollard, K. J., &Geissmann, F. (2010). Monocytes in atherosclerosis: Subsets and functions. *Nature Reviews Cardiology, 7*(2), 77–86. https://doi.org/10.1038/nrcardio.2009.228.

130. Yago, T., Fu, J., McDaniel, J. M., Miner, J. J., McEver, R. P., & Xia, L. (2010). Core 1-derived O-glycans are essential E-selectin ligands on neutrophils. *Proceedings of the National Academy of Sciences of the United States of America, 107*(20), 9204–9209.

131. Yago, T., Shao, B., Miner, J. J., Yao, L., Klopocki, A. G., Maeda, K., Coggeshall, K. M., &McEver, R. P. (2010). E-selectin engages PSGL-1 and CD44 through a common signaling pathway to induce integrin alphaLbeta2-mediated slow leukocyte rolling. *Blood, 116*(3), 485–494.

132. Yip, J., Shen, Y., Berndt, M. C., & Andrews, R. K. (2005). Primary platelet adhesion receptors. *IUBMB Life, 57*(2), 103–108.

133. Zarbock, A., & Ley, K. (2009). Neutrophil adhesion and activation under flow. *Microcirculation, 16*(1), 31–42.

134. Zarbock, A., Lowell, C. A., & Ley, K. (2007). Spleen tyrosine kinase Syk

is necessary for E-selectin-induced alpha(L)beta (2) integrin-mediated rolling on intercellular adhesion molecule-1. *Immunity, 26*(6), 773–783.

135. Zarbock, A., Polanowska-Grabowska, R. K., & Ley, K. (2007). Platelet-neutrophil-interactions: Linking hemostasis and inflammation. *Blood Reviews, 21*(2), 99–111.

136. Zarbock, A., Singbartl, K., & Ley, K. (2006). Complete reversal of acid-induced acute lung injury by blocking of platelet neutrophil aggregation. *Journal of Clinical Investigation, 116*(12), 3211–3219.

137. Zernecke, A., Bot, I., Djalali-Talab, Y., Shagdarsuren, E., Bidzhekov, K., Meiler, S., Krohn, R., Schober, A., Sperandio, M., Soehnlein, O., Bornemann, J., Tacke, F., Biessen, E. A., & Weber, C. (2008). Protective role of CXC receptor 4/CXC ligand 12 unveils the importance of neutrophils in atherosclerosis. *Circulation Research, 102*(2), 209–217. https://doi.org/10.1161/CIRCRESAHA.107.160697.

Chapter 3

WHITE BLOOD CELLS: OUR DEFENSE AGAINST INVADERS

"Avaunt! And quit my sight! Let the earth hide thee! Thy bone is marrowless. Thy blood is cold; thou hast no speculation in those eyes which thou dost glare with."
—Shakespeare, Macbeth

3.1. INTRODUCTION

With well-developed scientific understanding and advanced technologies, it is very clear that in the intricate world of the human body, a remarkable and well-specialized defense mechanism is operational 24/7, in order to safeguard us from unseen threats. This ever-vigilant defense mechanism operates silently against invaders and various types of abnormalities. Interestingly, these star pearls, named white blood cells, account for only 1% of the total cellular population, but their impact is big; especially in preserving our well-being, shielding us from infections, and orchestrating the symphony of immunity (Figure 3.1).

The current chapter explores the fascinating journey of white blood cells from their origin, multifaceted functions, and profound significance in both health and medicine. We will also discuss the remarkable advancements that continually expand our understanding and enable us to harness their incredible power.

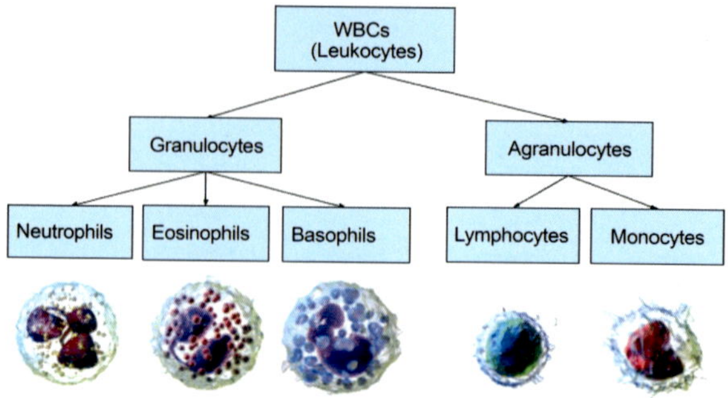

Figure 3.1. *An overview of white blood cells.*

Source: Ahmad R, Awais M, Kausar N, &Akram T, (2023). White blood cells classification using entropy-controlled deep features optimization. Diagnostics, 13(3), 352. https://doi.org/10.3390/diagnostics13030352: © 2023 by the authors. Licensee MDPI, Basel, Switzerland. This article is an open access article distributed under the terms and conditions of the Creative Commons Attribution (CC BY) license (https://creativecommons.org/licenses/by/4.0/).

3.2. THE ORIGIN AND DIVERSITY OF WHITE BLOOD CELLS

The human body is a very complex structure inside which a diverse range of processes are going on simultaneously, mostly regulated by trillions of cells working in a synchronized manner with different objectives. A specialized white blood cell is one such community always ready to fight off germs and invaders that can make us sick. Interestingly, many of you might be wondering where these brave cells originate from and whether they are only one type or if different forms also exist?

3.3. THE GENESIS OF IMMUNITY

The journey of white blood cells' origin and their transformation begins in the heart of our bones, in a place called the bone marrow. Bone marrow is soft, spongy tissue present inside our bones. It contains cells that are creators of blood cells and platelets. Bone marrow is the central place known for its brilliance in producing billions of cells present in our blood, functioning appropriately to make us what we are today!

All different types of blood cells are formed with the help of a process called "Hematopoiesis." Hematopoietic stem cells are the versatile cells present in the bone marrow and are blessed with the potency to develop into all blood cells, including cells of lymphoid and myeloid origin. Although these HSCs are present at multiple locations, including peripheral blood, bone marrow, and umbilical blood; the cells that originated in the marrow are identified to be the most accessible stem cells with the ability to produce multiple cells present in human blood.

The process of generating blood cells is remarkable and complex. These blood cells include white blood cells, also known as immune cells, along with red blood cells and platelets. Astonishingly, your body manufactures a staggering 100 billion cells every single day. This is essential because both immune cells and red blood cells have relatively short lifespans, and as the front-line warriors of your immune system, they are frequently sacrificed in the battle against everyday invaders.

In adults, hematopoiesis primarily unfolds within the bone marrow, the central cavity within your bones. The process kicks off with hematopoietic stem cells (HSCs), which are found in specialized bone marrow regions known as "niches." These HSCs possess the extraordinary ability to transform into any type of blood cell, a transformation finely regulated by growth factors and other mediators. As they progress through a sequence of steps that increasingly commit them to become specific cell types, HSCs ultimately give rise to a vast array of immune cells, such as monocytes, neutrophils, B cells, T cells, basophils, eosinophils – along with platelets and red blood cells. These newly formed cells are then released into your bloodstream. Hematopoiesis is a dynamic and finely tuned process that adjusts to meet your body's demands. For instance, during an infection, your body requires more immune cells, prompting an escalation in hematopoiesis. However, it should also be noted that uncontrolled hematopoiesis can lead to an excess of immune cells, resulting in heightened inflammation and the exacerbation of disease. Numerous unhealthy lifestyle choices and habits, especially those related to sleep, diet, and stress, can stimulate hematopoiesis and worsen immune and inflammatory conditions.

3.4. ORIGIN OF WHITE BLOOD CELLS (WBCS)

Once WBCs are produced in the bone marrow, young white blood cells are trained and matured in the bone marrow. You can imagine this process by considering bone marrow as a special school where young WBCs or immune cells go to learn their roles and develop unique skills. This "school" is incredibly important because it's where these cells prepare to become effective defenders of our bodies. Let's know how this process works (Figure 3.2):

- **Early Development:** When these immune cells are very young, they are a bit like students starting their education. They haven't fully decided what type of immune cells they will become yet.

- **Specialization:** As time goes on, these young cells start to specialize. It's like students picking a major in college. Some decide to become neutrophils, which are like the foot soldiers of the immune system. Others may choose to become B cells or T cells, which are like the generals, leading to specific parts of the immune response.

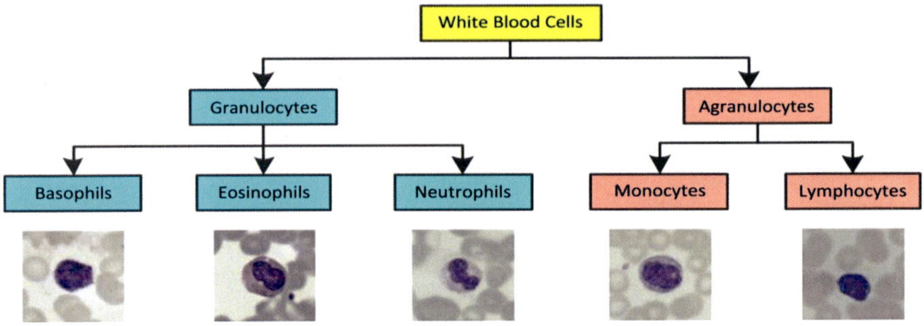

Figure 3.2.*The types of WBCs.*

Source: Chen, H., Liu, J., Hua, C. et al., (2022). Accurate classification of white blood cells by coupling pre-trained ResNet and DenseNet with SCAM mechanism. BMC Bioinformatics, 23, 282. https://doi.org/10.1186/s12859–022–04824–6: Open Access This article is licensed under a Creative Commons Attribution 4.0 International License, which permits use, sharing, adaptation, distribution and reproduction in any medium or format, as long as you give appropriate credit to the originalauthor(s) and the source, provide a link to the creative commons license, and indicate if changes were made. The images or other third-party material in this article are included in the article's Creative Commons license, unless indicated otherwise in a credit line to the material. If material is not included in the article's creative commons license and your intended use is not permitted by statutory regulation or exceeds the permitted use, you will need to obtain permission directly from the copyright holder. To view a copy of this license, visit http://creativecommons.org/licenses/by/4.0/. The Creative Commons Public Domain Dedication waiver (http://creativecommons.org/publicdomain/zero/1.0/) applies to the data made available in this article, unless otherwise stated in a credit line to the data.

- **Learning the Ropes:** Once they have chosen their path, these cells start to learn the specific skills they will need for their roles. Neutrophils, for example, become experts at chasing down and gobbling up bacteria, while B cells learn how to produce antibodies to tag invaders for destruction.

3.5. MATURATION PROCESS

As these young immune cells progress through their training in the bone marrow, they undergo a maturation process. This is a bit like students going through different grades in school, with each grade teaching them new things.

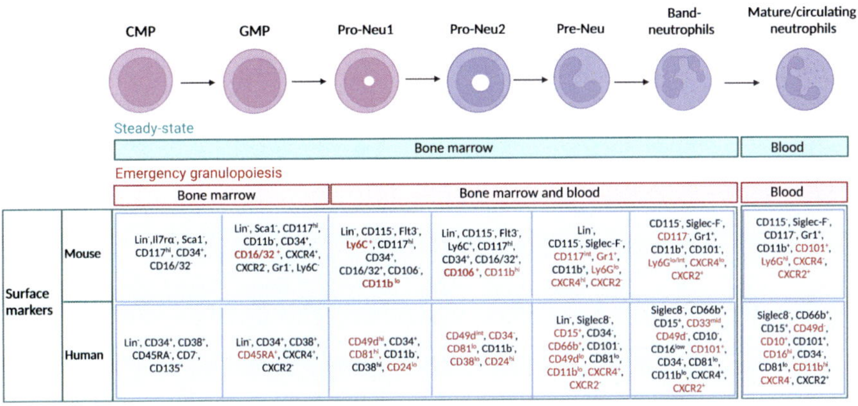

Figure 3.3. Key: BM: bone marrow; common myeloid progenitors (CMPs); granulocyte–monocyte progenitors (GMPps)

Source: Carnevale S, Di Ceglie I, Grieco G, Rigatelli A, Bonavita E, &Jaillon S, (2023) Neutrophil diversity in inflammation and cancer. Front. Immunol., 14, 1180810. doi: 10.3389/fimmu.2023.1180810:

- **Maturity Levels:** Cells become more and more mature as they advance in their training. This means they become better at what they do and more specialized in their roles.
- **Getting Ready for Action:** Think of this maturation process as preparing these cells for battle. They are equipped with all the tools they need to recognize and fight off invaders like bacteria, viruses, and other harmful substances.

Figure 3.3. As an example, the image below summarizes the stages in the maturation of a WBC (neutrophil here) in humans and mice.

- **Release Into Action:** Once they have completed their training and have reached the right level of maturity, these immune cells are ready to be sent out into the body. They leave the bone marrow and enter the bloodstream, where they can quickly respond to threats and help keep us healthy.

- Once young white blood cells are trained and matured for what they are destined to do, they are then released into the bloodstream from the bone marrow.

3.6. THE RELEASE INTO THE BLOOD STREAM

The process of releasing trained, mature white blood cells into the bloodstream is akin to these young defenders leaving school and heading into the world.

Here's how it happens:

- **Completion of Training:** After going through their maturation process in the bone marrow, which involves learning their roles and developing their skills, the WBCs are now fully prepared for action. They have become specialized in their roles, whether as neutrophils, B cells, T cells, or other types of immune cells.
- **Migration of Blood Vessels:** When it's time to be deployed, these mature WBCs move from the bone marrow to the blood vessels. Think of it as them packing their bags and heading to their new assignments.
- **Early Circulation:** Once inside the bloodstream, these WBCs are now on the move throughout your body. They travel along with the flow of blood, which carries them to different tissues, organs, and even areas where they are needed the most.
- **Patrolling and Defense:** As they circulate through your body, these white blood cells act like vigilant patrollers. They are on the lookout for any signs of trouble, such as invading bacteria, viruses, or other harmful substances. When they encounter a threat, they are quick to respond and initiate the immune defense mechanisms they have learned during their training.
- **Specific Roles:** Different types of white blood cells have specific roles. For instance, neutrophils are excellent at chasing down and engulfing bacteria, while B cells and T cells are like specialized commandos, orchestrating specific immune responses.
- **Adaptive Responses:** Some white blood cells, like B cells, also play a crucial role in adaptive immunity. They remember past encounters with certain invaders and can mount a faster and more targeted response if the same threat reappears.

Thus, the release of white blood cells into the bloodstream marks the moment when these skilled defenders are dispatched throughout your body. They patrol, detect, and respond to threats, helping to keep you healthy by defending against infectious and other dangers. This orchestrated process ensures that your immune system is always ready to protect you. In fact, several hormones and signals have been documented to have an effect on the migration of leukocytes. Table 3.1 summarizes these aspects presented below as documented by Ince et al. (2019).

Table 3.1. The impact of signals on the migration of leukocytes (please note that the references in numbers have been compiled by the authors Ince et al, 2019; interested readers may read this review to access these references)

	Duration of stimulus	Receptor	Compound	Cell	Effect	References
Catecholamines	Acute (2 h in vivo)	Not assessed	A	Rat CD62L neg. monocytes	Increased numbers in blood	(20)
	Acute (2 h in vivo)	Not assessed	A, NA, NA+A	Rat monocytes	Increased numbers in blood, decreased CD62L expression	(20)
	Acute (2 h in vivo)	Not assessed	A, NA, NA+A	Rat CD62L neg. neutrophils	Increased numbers in blood	(20)
	Acute (2 h in vivo)	Not assessed	A, NA, NA+A	Rat CD62L pos. neutrophils	Increased numbers in blood	(20)
	Acute (2 h in vivo)	Not assessed	A	Rat CD62L neg. T, NK cells	Decreased numbers in blood	(20)
	Acute (2 h in vivo)	Not assessed	NA+A	Rat CD62L pos T, NK cells	Decreased numbers in blood	(20)
	Acute (2 h in vivo)	Not assessed	A, NA, NA+A	Rat NK cells	Decreased CD62L expression	(20)
	Acute (2 h in vivo)	Not assessed	A, NA+A	rat lymphocytes	Decreased numbers in blood, decreased CD62L expression	(20)
	Acute (2 h in vivo)	Not assessed	A, NA+A	Rat cytotoxic T cells	Decreased numbers in blood	(20)
	Acute (2 h in vivo)	Not assessed	NA	Rat CD62L neg. B cells	Decreased numbers in blood	(20)
	Acute (2 h in vivo)	Not assessed	A, NA, NA+A	Rat CD62L pos. B cells	Decreased numbers in blood, CD62L expression unaffected	(20)

	Acute (2 h in vivo)	Not assessed	NA	Rat B cells	Decreased numbers in blood	(20)
	Acute (2 h in vivo)	Not assessed	A, NA+A	Rat B cells	Decreased numbers in blood	(20)
	Acute (4 h in vitro)	Not assessed	A	Human neutrophils and monocytes	Increased CD11b expression; suppression of LPS-induced CD11b and CD18 expression	(4)
	Acute (4 h in vitro)	not assessed	A	Human white blood cells	Dose-dependent increase in IL-8 levels; suppression of LPS-induced production of IL-1β, IL-8, and CCL2	(4)
	Acute (90 min in vitro)	β-AR	A, NA, Isoprenaline	Human PMNs	Reduced fMLP-induced migration, CD11b/CD18 expression and ROS production	(21)
	Acute (30 min pre-treatment in vitro)	Not assessed	A	Human neutrophils	Reduced adhesion to HUVECs by down-modulation of EC adhesion molecule expression	(6)
	Acute (30 min in vitro)	Not assessed	A, NA	Mouse macrophages/ neutrophils	Dose-dependent activation of NFκB, decrease of IκBα levels	(9)
	Acute (4 h in vitro)	Not assessed	A, NA	Mouse macrophages	Dose-dependent activation of NFκB, release of TNFα, IL-1β, IL-6, CXCL2	(9)
	Chronic (8 days in vivo)	β2-AR	A	Mouse macrophages	Production of IL-6, leading to persistent neutrophil trafficking	(22)
	None (endogenous)	β2-AR	Endogenous	Human/ mouse macrophages	Changes in VCAM-1 expression levels	(8)

	None (endogenous)	β2-AR	Endogenous	Mouse lymphocytes	Inhibition of egress from lymph node through CCR7 and CXCR4	(10, 11)
	Acute (20 min in vitro)	α2-AR	Xylazine, UK14304	Human neutrophils	Reduced trafficking without affecting CD62L and CD11b expression	(5)
	Acute (6 h in vitro)	α2-AR	Xylazine, UK14304	Human endothelial cells	Decreased transendothelial migration of neutrophils	(5)
	Chronic (5 days in vivo)	β3-AR	BRL37344	Mouse endothelial cells	Upregulation of VCAM-1, P- and E-selectin expression, more BM homing	(7)
Glucocorticoids	Acute (6 h in vivo)	GR	Dexamethasone	Human granulocytes	Increased numbers in blood; detached more easily in ex vivo assay	(3)
	Acute (2 h in vitro)	GR	Dexamethasone	Human granulocytes	Detached more easily in in vitro assay	(3)
	Chronic (7 days in vivo)	GR	Mifepristone (RU486)*	Rat neutrophils	Increased numbers in blood; CD62L expression increased in blood, decreased in BM	(13)
	Int. (24 h and 2 h in vivo)	GR	Mifepristone (RU486)*	Mouse neutrophils	Decreased annexin A1, altered neutrophil maturation and homing	(12)
	Acute (10 h in vivo)	GR	Mifepristone (RU486)*	Human T cells	Increased CXCR4 expression in behavioral rest phase, decreased in active phase (inverse to blood numbers)	(15)

	None (endogenous)	GR	Endogenous	Mouse T cells	When T cell GR is disrupted, CXCR4 expression is reduced and homing impaired in active phase	(23)
	Acute (8 h in vivo)	MR	Fludrocortisone	Human naïve T cells	Agonism decreased circulating numbers, increased CXCR4 expression (in vivo)	(24)
	Acute (2–4 h in vitro)	MR	Spironolactone*/ Fludrocortisone	Human naïve T cells	Agonism increased CXCR4 and CD62L expression, antagonism decreased CD62L and CCR7 expression	(24)
	Acute (1 h pre-treatment in vivo)	GR	Dexamethasone	Mouse leukocytes	Reduced LPS-induced adhesion	(14)
	Int. (18 h and 1 h pre-treatment in vivo)	GR	Mifepristone (RU486)*	Mouse leukocytes	Increased LPS-induced adhesion, but reduced endothelial VCAM-1 expression	(14)
	Acute (1 h pre-treatment in vivo)	GR	Dexamethasone	Mouse neutrophils	Inhibited LPS-induced neutrophil migration into lungs if administered during rest phase, but not during active phase	(16, 17)
	Chronic (trait assessments)	Not assessed	Endogenous	Macaque leukocytes	Positive correlation of cortisol and neutrophil numbers in blood in low-nervous animals, no association in high nervous animals	(25)

	Chronic (16 months) + acute (2 h)	GR	Endogenous (stress) + dexamethasone	Macaque leukocytes	Stressed animals show reduced sensitivity to dexamethasone-induced reduction of circulating lymphocytes	(26)

Source: Ince LM, Weber J &Scheiermann C (2019) Control of Leukocyte Trafficking by Stress-Associated Hormones. Front. Immunol., 9, 3143. doi: 10.3389/fimmu.2018.03143:

3.7. THE FRONTLINE AND BEYOND

From the bloodstream, the white blood cells travel to various tissues and organs, where they await their call to action. We will discover how they are strategically positioned throughout the body to respond to infections and maintain our health.

Conclusively, the birthplace of white blood cells is a miraculous process that takes place within our bones. From hematopoiesis to differentiation, maturation, and finally, release into the bloodstream. Further to this, understanding their origin and creation allows us to appreciate the vital role they play in safeguarding our health and keeping us well.

3.8. A MULTITUDE OF WARRIORS

Now that we have learned how white blood cells originate in bone marrow, and how well synchronized the process of their origin, training, and maturation is; it is interesting to note that white blood cells are commonly known as the "multitude of warriors," and their specific role in detail.

3.8.1. Neutrophils: The Frontline Defenders

Neutrophils are the type of white blood cells (WBC) that originate in the bone marrow. They are among the first immune cells to be produced in response to an infection or injury. Furthermore, they are the most abundant WBCs present in the bloodstream, making up about 60–70% of all WBCs (Nauseef&Borregaard, 2014). Neutrophils, sometimes called polymorphonuclear (PMN) leukocytes, are the most abundant type of cell in the human bloodstream. They are churned out in large quantities by the bone marrow, producing approximately 100 billion of them daily (Scapini&Cassatella, 2014). Normally, neutrophils circulate in the bloodstream, travel to tissues, perform their functions, and are eventually cleared away by macrophages, all within the span of a single day (Figure 3.4).

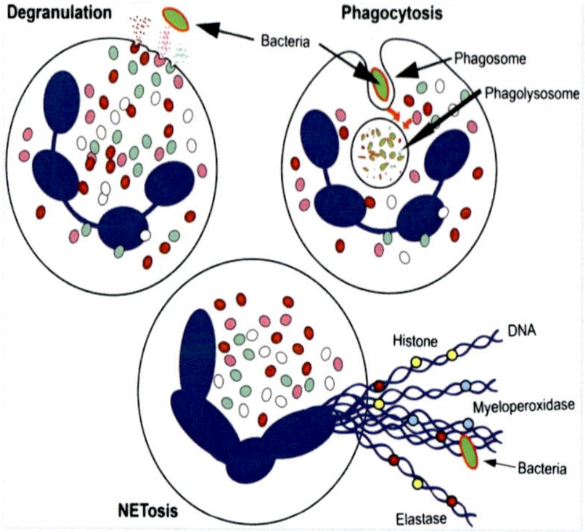

Figure 3.4.*A few aspects of neutrophils.*

Source: Rosales C, (2018) Neutrophil: A cell with many roles in inflammation or several cell types? Front. Physiol., 9, 113. doi: 10.3389/fphys.2018.00113: Copyright © 2018 Rosales. This is an open-access article distributed under the terms of the Creative Commons Attribution License (CC BY). The use, distribution or reproduction in other forums is permitted, provided the original author(s) and the copyright owner are credited and that the original publication in this journal is cited, in accordance with accepted academic practice. No use, distribution or reproduction is permitted which does not comply with these terms.

Neutrophils play crucial roles in the innate branch of our immune system, constantly patrolling the body to detect signs of microbial infections. When they detect such invaders, they respond swiftly by capturing and neutralizing these pathogens. Neutrophils have three primary ways of fighting against the infection i.e., phagocytosis, degranulation, and forming neutrophil extracellular traps (NET) that are composed of expelled nuclear material that ensnares pathogens (Tecchio&Cassatella, 2016).

Traditionally, neutrophils were thought to serve only these antimicrobial functions. However, recent research across various fields of neutrophil biology has revealed a much broader spectrum of their roles (Greenlee-Wacker, 2016). Neutrophils respond to multiple signals and react by producing various cytokines (signaling molecules) and other inflammatory factors. These substances not only influence and regulate inflammation but also impact the immune system. Today, we understand that neutrophils are complex and active cells at the genetic level, capable of generating cytokines, influencing neighboring cells, and even contributing to the resolution of inflammation. They also play roles in long-term responses, interact with diseases, including cancer, and have a part in the innate immune response memory. This expanded understanding of neutrophils has illuminated their significance far beyond their initial role as microbial killers (Netea et al., 2016).

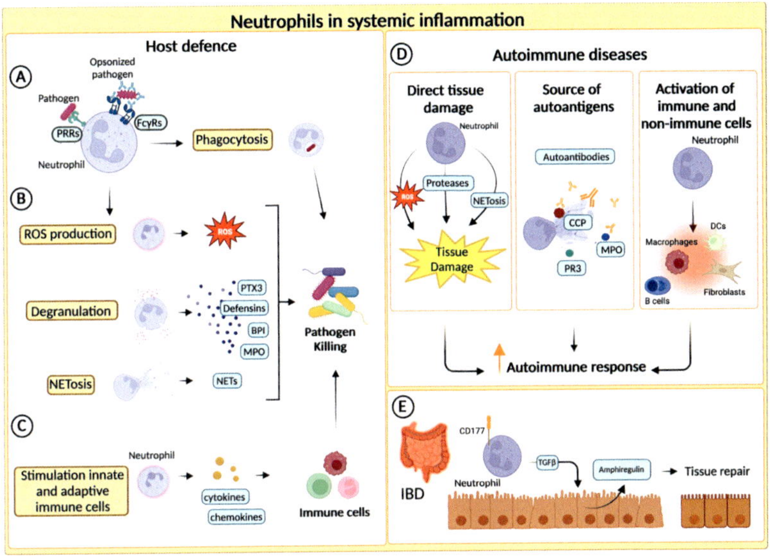

Figure 3.5. *The role of neutrophils in inflammation.*

Source: Carnevale S, Di Ceglie I, Grieco G, Rigatelli A, Bonavita E &&Jaillon S, (2023). Neutrophil diversity in inflammation and cancer. Front. Immunol., 14, 1180810. doi: 10.3389/fimmu.2023.1180810:

Some of the essential characteristics of neutrophils are:

- They are often referred to as the "First responders" because they are incredibly fast in reaching the site of infection or injury. They can quickly migrate from the bloodstream to the affected tissue (Nauseef&Borregaard, 2014).

- Their primary role is phagocytosis, a process in which they engulf and digest invading microorganisms like bacteria, fungi, and other pathogens. This is a very crucial defense mechanism against invading infection.
- Neutrophils are guided to the site of infection by a process called "Chemotaxis." The chemical signals released by damaged tissues or by other immune cells attract neutrophils to the site of the threat.

Figure 3.5 shows the role of neutrophils in inflammation has been extensively compiled by researchers.

- Neutrophils employ a respiratory burst, which is a burst of oxygen consumption and production of reactive oxygen species (ROS) during phagocytosis. This burst of oxidative activity helps in killing and digesting the engulfed microorganisms.
- Neutrophils have a relatively short lifespan, typically living for only a few hours to a couple of days. After performing their phagocytic duties, they often undergo apoptosis (Programmed cell death) and are then cleared from the body.
- Neutrophils can also release the NETs, i.e., neutrophil extracellular traps, which are web-like structures made up of DNA, histones, and antimicrobial proteins. NETs trap and immobilize pathogens, enhancing the immune response (Nauseef&Borregaard, 2014).
- Moreover, neutrophils play a significant role in the inflammatory response. Their arrival at the site of infection can lead to the classic signs of inflammation, including redness, swelling, and heat as they work to eliminate the threat.
- Neutrophils are part of a coordinated immune response. They interact with other immune cells like macrophages, dendritic cells, and lymphocytes to ensure a comprehensive defense strategy.

Furthermore, as an interesting case, the functioning of neutrophils in both tumor promotion and in the fight against tumors has been presented Figures 3.6 and 3.7.

Thus, neutrophils are frontline defenders with a rapid response to infections and injuries. They excel at phagocytosis, chemotaxis, and the respiratory bursts, all of which help them efficiently eliminate invading microorganisms. While they have a short lifespan, their role in inflammation and their ability to cooperate with other immune cells make them indispensable components of the body's immune defense system.

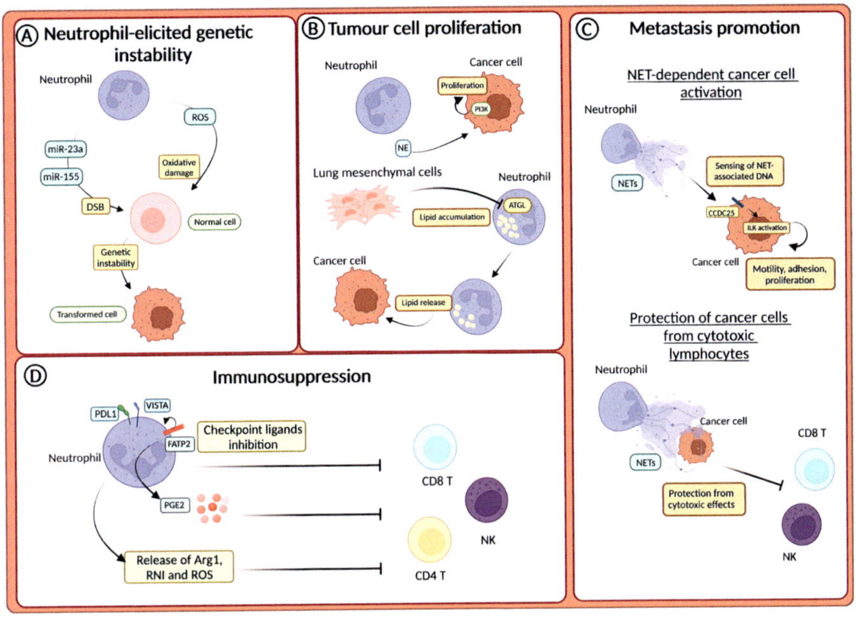

Figure 3.6. The functioning of neutrophils in both tumor promotion and in the fight against tumors.

Key: (A) Neutrophil-derived reactive oxygen species (ROS) or miR-23a and miR-155 promote double-strand breaks (DSBs). (B) Neutrophils sustain tumor proliferation via neutrophil elastase (NE) and subsequent activation of PI3K. Inhibition of adipose triglyceride lipase (ATGL) causes lipids toaccumulate, which are then released to feed tumor cells. (C) CCDC25 on tumor cells is a sensor for neutrophil extracellular trap (NET)-associated DNA, which activates integrin-linked kinase (ILK) signaling. (D) Neutrophils cause immunosuppression via lymphocyte checkpoints (e.g., PD-L1 and VISTA), the production of ROS, arginase 1 (Arg1) or reactive nitrogen intermediates (RNI), and fatty acid transporter protein 2 (FATP2)-dependent production of prostaglandin E2 (PGE2).

Source: Carnevale S, Di Ceglie I, Grieco G, Rigatelli A, Bonavita E, &Jaillon S, (2023). Neutrophil diversity in inflammation and cancer. Front. Immunol., 14, 1180810. doi: 10.3389/fimmu.2023.1180810:

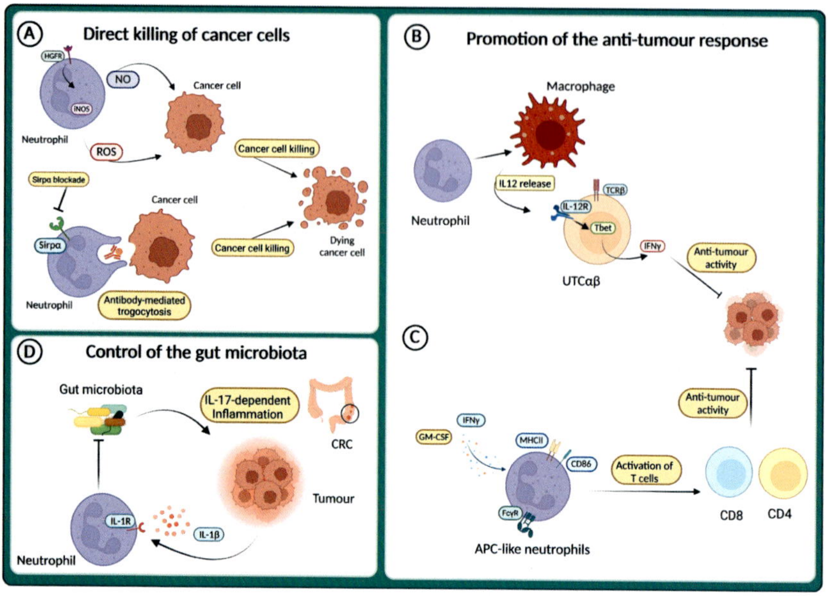

Figure 3.7. The functioning of neutrophils in both tumor promotion and in the fight against tumors.

Key: (A) Killing of cancer by the release of reactive oxygen species (ROS) and NO or via antibody-mediated reactions. (B) Neutrophils help IL-12 production by macrophages for αβ T cells (UTCαβ) to produce IFNγ and suppress tumors. (C) Induction of APC-like neutrophils by GM-CSF and IFNγ, or by FcγR-mediated endocytosis of antibody-antigen to activate T cell anti-tumor response. (D) Limiting gut microbiota and dampening microbiota-dependent inflammation in colorectal cancer (CRC)

Source: Carnevale S, Di Ceglie I, Grieco G, Rigatelli A, Bonavita E &Jaillon S (2023) Neutrophil diversity in inflammation and cancer. Front. Immunol., 14, 1180810. doi: 10.3389/fimmu.2023.1180810:

3.8.2. Lymphocytes: The Architects of Immunity

Lymphocytes are the type of white blood cells (WBCs) that are formed in the bone marrow, similar to other WBCs. However, their maturation and specialization occur mainly in the lymphoid tissues. There are mainly two main types of lymphocytes, namely B cells and T cells. They have a distinct role in the immune system (Summers et al., 2010). Key aspects of these cells are documented in Figure 3.8.

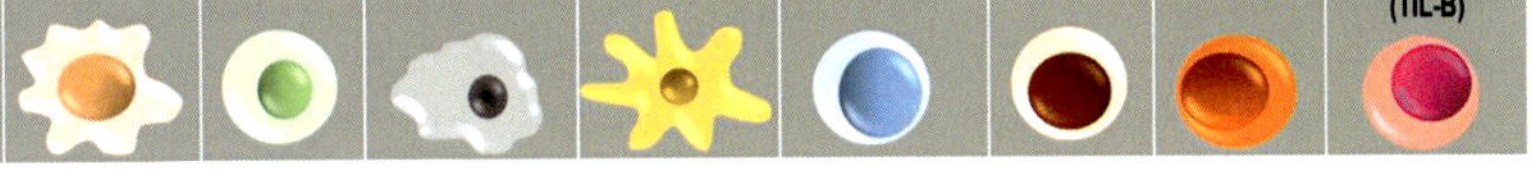

Immune cell type	M1/M2 Macrophages	N1/N2 Neutrophils	Myeloid-Derived Suppressor cells	Dendritic cells (DCs)	Natural killer cells (NKs)	CD4⁺ T cells	Regulatory T cells (Treg)	Tumor infiltrating B lymphocytes (TIL-B)
	M1	N1				Th1		
Anti-tumorigenic functions	Antigen presentation to T cells	Tumor rejection and cytotoxicity		Antigen presentation and costimulatory signals for T cell activation	Cytotoxic immune response (Fas/FasL)	Recruitment of CTLs, NK and M1 macrophages		Antigen presentation to T cells
	Secretion of IL-1β, IL-6, IL-12, IL-23, TNF-α, IFN-γ, CXCL10, CXCL12, ROS, NO and iNOS	Secretion of FAS, TNF-α, CCL3, ICAM-1		Secretion of TNF-α, granzyme B, lymphotoxin, sTRAIL	Secretion of IFN-γ, granzyme B , perforin	Secretion of IL-2, TNF-α, IFN-γ		Secretion of TNF-α, IFN-γ, IL-2, IL-4. IL-6, IL-12, IL-13, lymphotoxin
	M2	N2				Th2/Th17		
Pro-tumorigenic functions	Immune suppression (CTLA-4, PD-1)	Immune suppression (PD-L1)	Immune suppression (PD-L1)	Immune suppression (TIM-3, PD-1, PD-L1)	Immune suppression (KIRs, PD-1, CTLA-4, TIM-3, LAG-3)	Promoting tumor growth and angiogenesis	Immune suppression (CTLA-4,PD-1 LAG-3)	Immune suppression (PD-L1)
	Promoting angiogenesis and metastasis	Promoting tumor growth and angiogenesis	Promoting angiogenesis				Suppression of APCs and effector T cells	Promoting tumor growth and angiogenesis
	Secretion of TNF-β, IL-1β, IL-10, CCL5, CCL17, CCL18, CCL22, ARG1, MMP-9	Secretion of CXCR4, CCL2, CCL5, VEGF, MMP-9, ARG1	Secretion of IL-10, TGF-β, CCL2, CCL4, CCL5, VEGF, ARG1, IDO, MMP-9	Secretion of IL-10, TGF-β, VEGF, ARG2, iNOS, IDO and β-catenin	Secretion of IL-4, IL-10, IL-13 TGF-β, VEGF, CCL3, CCL4	Secretion of IL-4, IL-5, Il-6, IL-10, IL-13, IL-25 (Th2); IL-21, Il-17, IL-22 (Th17)	Secretion of IL-10, Il-35, TGF-β, IDO, granzymes perforin, adenosine	Secretion of IL-10, IL-35 and TGF-β
References	[12-18]	[19-23]	[24-28]	[29-35]	[36, 37]	[39, 42]	[43-45]	[46-53]

Figure 3.8. *Two main types of lymphocytes, namely B cells and T cells.*

Source: Capitani N, Patrussi L, &Baldari CT, (2021).Nature vs. nurture: The two opposing behaviors of cytotoxic T lymphocytes in the tumor microenvironment. International journal of molecular sciences, 22(20), 11221. https://doi.org/10.3390/ijms222011221: © 2021 by the authors. Licensee MDPI, Basel, Switzerland. This article is an open access article distributed under the terms and conditions of the Creative Commons Attribution (CC BY) license (https://creativecommons.org/licenses/by/4.0/).

- B cells, or B lymphocytes, are responsible for humoral immunity. They mature in the bone marrow and are equipped with antibodies (Immunoglobulins) on their cell surfaces.
- When B cells encounter an antigen (a foreign substance), they can differentiate into plasma cells, which produce antibodies. Antibodies play a crucial role in tagging invaders like bacteria or viruses for destruction and removal by other immune cells.
- T cells or T lymphocytes mature in the thymus gland. They are

involved in cell-mediated immunity, which means they directly attack infected or abnormal cells in the body.

- There are different types of cells including cytotoxic T cells, i.e., CD8+, Helper T cells (CD4+), and regulatory T cells (Tregs). Each has a specific role in the immune responses.
- Recognition of Antigens is a primary function of T cells, which carry receptors on their surfaces. These receptors are highly diverse, allowing lymphocytes to identify a wide range of invaders.
- After encountering an antigen, some lymphocytes become memory cells. These cells "remember" the specific antigens and lead to a faster and more robust immune response if the same antigen is encountered in the future.
- Lymphocytes also play a central role in adaptive immunity, a type of immune response that adapts and improves with each encounter with a pathogen. They provide the body with long-lasting protection against previously encountered threats (Figure 3.9).

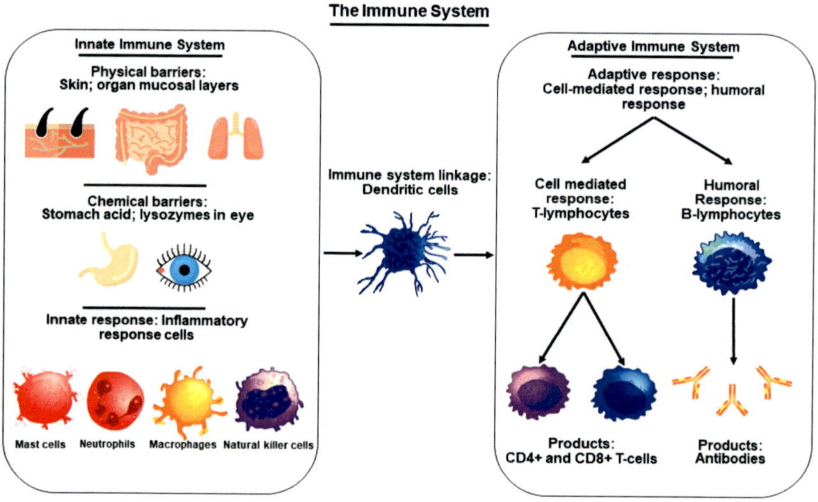

Figure 3.9. *Vital aspects of the immune system.*

Source: Manna PR, Gray ZC, & Reddy PH, (2022). Healthy immunity on preventive medicine for combating COVID-19. Nutrients, 14(5):1004. https://doi.org/10.3390/nu14051004: © 2022 by the authors. Licensee MDPI, Basel, Switzerland. This article is an open access article distributed under the terms and conditions of the Creative Commons Attribution (CC BY) license (https://creativecommons.org/licenses/by/4.0/).

- Lymphocytes are responsible for immune memory, which is the ability of the immune system to remember past infections. This memory allows for quicker and more effective responses upon re-exposure to the same pathogen (Nauseef&Borregaard, 2014).
- All lymphocytes work in coordination with other immune cells like antigen-presenting cells (APCs), such as dendritic cells, to initiate and regulate immune responses. Helper T cells, in particular, play a central role in orchestrating immune reactions.
- It is very important to note that lymphocytes can turn out to be very harmful if they are programmed in a different manner; autoimmunity is one such crucial example, depicting their harm. Sometimes, lymphocytes can mistakenly target cells of our own body, leading to autoimmune diseases. In allergies, they may overreact to harmless substances, causing allergic reactions.

Thus, lymphocytes are essential components of our immunity and are considered to be the building blocks of a person's immune system. They play a crucial role in recognizing foreign antigens, forming immune memory, and coordinating immune responses. Their ability to remember past encounters with pathogens is crucial for the long-term protection against diseases (Adrover et al., 2016).

3.8.3. Monocytes and Macrophages: The Clean-Up Crew

Monocytes are the kind of white blood cells that originate in the bone marrow from hematopoietic stem cells, the same as with other types of WBCs. Once they are produced, these monocytes are released into the bloodstream and circulate throughout our body. Once in the bloodstream, they perform the following functions (Martin et al., 2003). Interesting aspects of these cells are presented below:

- These cells are the essential components of the innate immune responses and serve as vigilant sentinels, patrolling the bloodstream and tissues, constantly on the lookout for signs of infections, injury, and inflammation.
- The primary role of monocytes and their mature counterparts known as macrophages is phagocytosis. This is the process by which they engulf and digest foreign invaders, including bacteria, viruses, and other cellular debris.
- When monocytes detect an infection or damage in tissues, they can leave the bloodstream and enter the affected area (Nauseef&Borregaard, 2014).

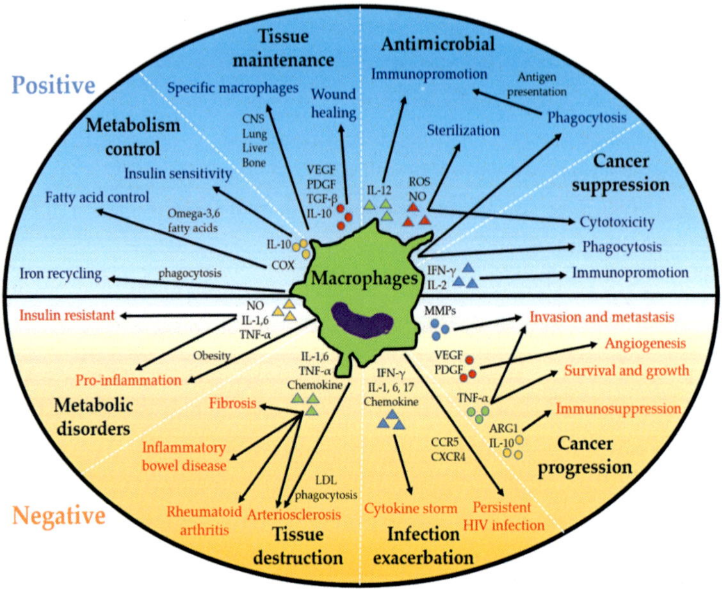

Figure 3.10. *Research is unraveling the myriad roles of these cells.*

Key: IFN, interferon; IL, interleukin; ROS, reactive oxygen species; NO, nitric oxide; VEGF, vascular endothelial growth factor; PDGF, platelet-derived growth factor; COX, cyclooxygenase; MMPs, matrix metalloprotease; TNF, tumor necrosis factor; ARG, arginase; CCR, CC chemokine receptor; CXCR, CXC chemokine receptor; LDL, low-density lipoprotein.

Source: Kadomoto S, Izumi K, &Mizokami A, (2022). Macrophage polarity and disease control. International Journal of Molecular Sciences, 23(1), 144. https://doi.org/10.3390/ijms23010144: © 2021 by the authors. Licensee MDPI, Basel, Switzerland. This article is an open access article distributed under the terms and conditions of the Creative Commons Attribution (CC BY) license (https://creativecommons.org/licenses/by/4.0/).

- Once in affected tissues, monocytes undergo a major transformation into macrophages. Macrophages are larger, more specialized phagocytic cells with enhanced abilities to clear pathogens and cellular debris.
- In addition to phagocytosis, these macrophages are essentially used for tissue repair and regeneration. Studies have confirmed that macrophages can release certain types of growth factors that promote the healing process along with certain types of anti-inflammatory cytokines, which are responsible for reducing inflammation drastically.

- Additionally, macrophages can also act as key players in coordinating immune responses. They present antigens (pieces of pathogens) to other types of immune cells, particularly T cells, to initiate specific immune responses.
- After an infection is cleared, macrophages remain active in the clean-up process, wherein dead cells and other cellular debris are removed to restore tissue health and regenerate the same.
- Moreover, monocytes and macrophages are involved in the inflammatory immune responses by releasing inflammatory mediators that recruit other immune cells to the site of injury.
- Dysregulation of monocytes and macrophage activity can contribute to chronic inflammatory diseases and conditions like atherosclerosis, arthritis, and certain kinds of autoimmune diseases.

Figure 3.10. Research is unraveling the myriad roles of these cells; the Figure 3.10 summarizes these aspects beautifully.

Thus, these clean-up crews are responsible for reducing the inflammation and minimizing the infection by conducting the process of phagocytosis to kill the bacteria and viruses and cleaning up of the pathogens along with other cellular debris. Further to this, these monocytes along with macrophages play a crucial role in maintaining tissue health and regenerating the same (Eash et al, 2010).

3.8.4. Eosinophils and Basophils: Specialized Guardians

Eosinophils, like every other white blood cell, are generated in the bone marrow through a process called hematopoiesis with the help of highly potent hematopoietic stem cells. Once formed, trained and matured eosinophils are released into the bloodstream for further specific actions (Zhang et al., 2015). Their main functions can be noted as:

- Eosinophils are critically involved in defending our bodies against parasitic infections, such as helminths, also commonly known as body worms (Nauseef&Borregaard, 2014).
- They are also involved in modulating allergic reactions and asthma. They release certain types of anti-inflammatory cytokines that contribute to the allergic responses and help control inflammation during these conditions.
- Eosinophils can phagocytose parasites, bacteria, and other cellular components, the same as their other counterparts discussed above. However, it should be noted that eosinophils are especially important for attacking and cleaning helminths.

- These cells can also migrate to the site of injury or inflammation in response to the chemical stimuli sent by the ailing cells.
- These cells are packed with specific granules containing certain important enzymes, toxins, and proteins that help in attacking and destroying parasites.

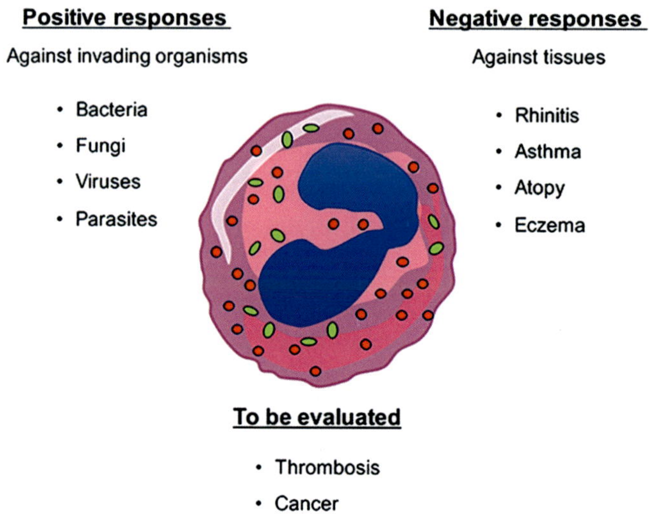

Figure 3.11. *The roles of eosinophils.*

Source: Ferrari D, Vuerich M, Casciano F, Longhi MS, Melloni E, Secchiero P, Zech A, Robson SC, Müller T, &Idzko M, (2020) Eosinophils and purinergic signaling in health and disease. Front. Immunol., 11, 1339. doi: 10.3389/fimmu.2020.01339: Copyright © 2020. This is an open-access article distributed under the terms of the Creative Commons Attribution License (CC BY). The use, distribution or reproduction in other forums is permitted, provided the original author(s) and the copyright owner(s) are credited and that the original publication in this journal is cited, in accordance with accepted academic practice. No use, distribution or reproduction is permitted which does not comply with these terms.

Similarly, basophils also originate from the same source with the help of HSCs. However, they are less common granulocytes compared to eosinophils and make up a very small percentage of total white blood cells. These cells play a pivotal role in subsiding immediate hypersensitive allergic reactions by releasing histamine and other inflammatory substances that are responsible for many allergic symptoms like itching, swelling, and increased mucus production (Massena et al., 2015).

Figure 3.12. *The wide range of implications of these cells have been depicted by Miyake et al. (2022).*

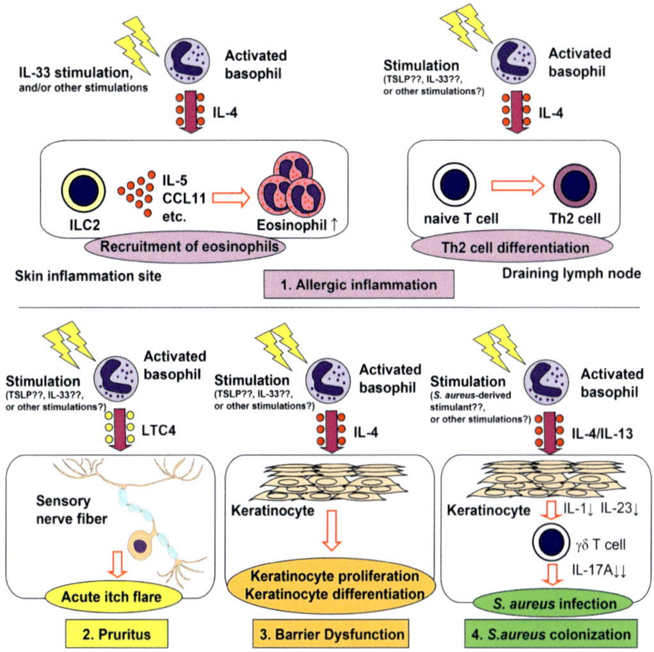

Figure 3.12.*The wide range of implications of these cells.*

Source: Miyake K, Ito J, &Karasuyama H, (2022) Role of Basophils in a broad spectrum of disorders. Front. Immunol., 13, 902494. doi: 10.3389/fimmu.2022.902494:

3.9. THE SENTINEL ROLE OF WHITE BLOOD CELLS

Imagine having ever-watchful security around your house, patrolling 24/7! In the same way, white blood cells do for your body. These ready-to-detect vigilant crew are guarding us by detecting and neutralizing threats before they can harm us in any way! As already discussed in our previous sections, white blood cells perform many different functions, which will be explained in detail in subsequent paragraphs.

3.10. THE VIGILANT WATCHMEN OF THE BLOOD STREAM

The primary functions of the white blood cells can be noted as:

The white blood cells, also known as leukocytes, act as sentinels including neutrophils, monocytes (which mature into macrophages), and dendritic cells. These cells are strategically positioned throughout our body. The role and function of each of these cells is explained in detail earlier (Netea et al, 2016).

3.11. THE MOBILIZATION AND THE PRECISION TARGETING

Imagine your bloodstream as a bustling highway network, with white blood cells as the vigilant traffic patrol. When a threat is detected, these WBCs mobilize swiftly and with precision to neutralize it. Here's how it all happens (Zhang et al, 2015):

- **Detection of Threat:** The journey begins when your body detects a potential threat, like invading bacteria or viruses. Other cells, known as antigen-presenting cells (APCs), play a crucial role in recognizing these threats and sounding the alarm (Woodfin et al, 2011).
- **Chemical Signaling:** When a threat is identified, cells release signaling molecules called cytokines. These serve as emergency beacons, guiding WBCs to the trouble spot. It is just like sending out a distress call to summon the troops.
- **Selecting The Right Soldier:** Different types of WBCs are specialized for various tasks. Neutrophils are, for example, like those foot soldiers, excellent at chasing down and engulfing bacteria. T cells and B cells are more like specialized commandos, orchestrating targeted immune cells. Monocytes and macrophages act as the clean-up crew, clearing away the cellular debris.
- **Mobilization From Reservoirs:** WBCs are strategically positioned in different parts of your body. Some are stored in your bone marrow, others in your lymph nodes, and some even patrol your tissues. When the alarm is raised, they are released from these reservoirs and sent into action.
- **Chemotaxis:** WBCs follow chemical gradients, moving in the

direction of higher cytokine concentrations. It is as if they can sense trouble and are drawn towards it. This type of movement is known as chemotaxis. Chemotaxis ensures that WBCs arrive precisely at the particular location whenever and wherever required (Woodfin et al, 2011).

- **Diapedesis:** To reach the site of infection or injury, WBCs need to leave the bloodstream and enter the affected tissue. They do this through a process called diapedesis. It's like crossing a security checkpoint to get to the scene of the emergency.

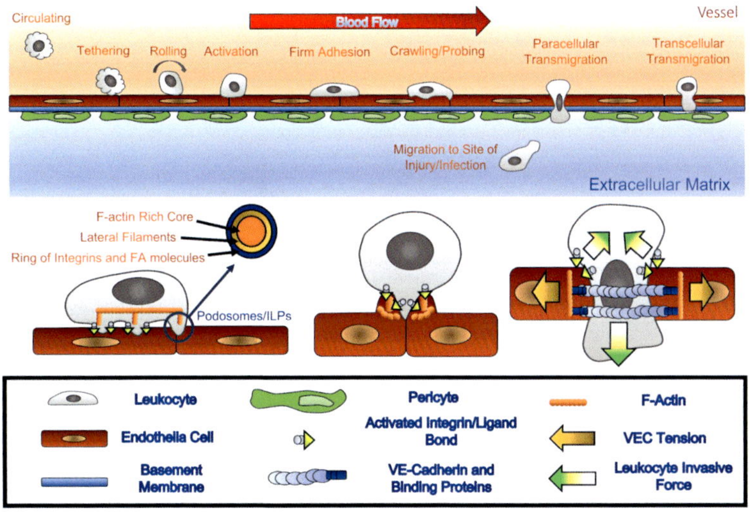

Figure 3.13. *The process of diapedesis.*

Source: Schwartz AB, Campos OA, Criado-Hidalgo E, Chien S, del Álamo JC, Lasheras JC, & Yeh YT, (2021) Elucidating the biomechanics of leukocyte transendothelial migration by quantitative imaging. Front. Cell Dev. Biol., 9, 635263. doi: 10.3389/fcell.2021.635263:

- **Engulfing the Enemy:** Once at the site, WBCs engage the enemy. Neutrophils, for instance, use a process called phagocytosis to engulf and digest invading bacteria. Other cells, like T cells and B cells, coordinate immune responses by recognizing specific antigens on the threat's surface.

- **Communication:** WBCs also communicate with each other. They release chemical signals to coordinate their actions and recruit more troops if needed. This communication ensures a coordinated response.

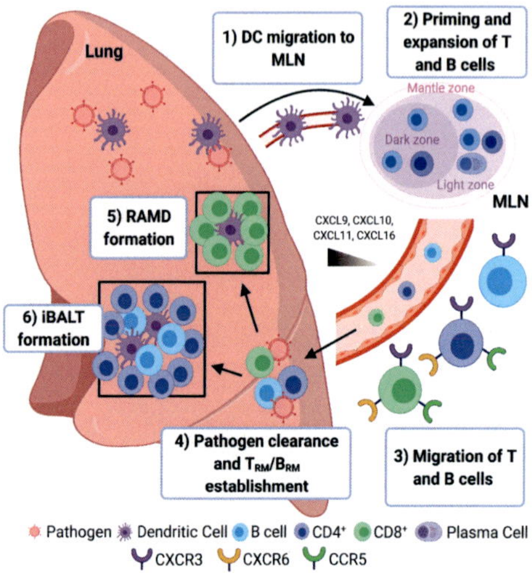

Figure 3.14. *The development of resident memory cells in the lung.*

Key: dendritic cells (DCs), mediastinal lymph node (MLN), germinal Centres (GCs), pulmonary-resident memory T cells (TRM), IgM+ pulmonary-resident memory B cells (BRM), repair-associated memory depots (RAMD), inducible bronchus-associated lymphoid tissue (iBALT)

Source: Humphries DC, O'Connor RA, Larocque D, Chabaud-Riou M, Dhaliwal K and Pavot V, (2021) Pulmonary-resident memory lymphocytes: pivotal orchestrators of local immunity against respiratory infections.Front. Immunol. 12, 738955. doi: 10.3389/fimmu.2021.738955:

- **Resolution and Clean Up:** After the threats are neutralized, some WBCs switch to clean-up mode. Monocytes, for example, can transform into macrophages that clear away dead cells and debris.

- **Immunological Memory:** Some WBCs, like memory T cells, "remember" past encounters with threats. This memory allows your immune system to respond more rapidly and effectively if the same threat returns (Netea et al., 2016).

As an example, the development of resident memory cells in the lung has been depicted Figure 3.14.

3.12. THE ARSENAL OF WHITE BLOOD CELLS

As discussed already, WBCs are armed with a complete set of armies. Their coordinated efforts, specialized roles, and diverse weaponry make them a formidable force against infections, inflammation, and damage. Let's take a deep dive to understand the basic overview of the immune response.

3.12.1. Phagocytosis: Engulfing the Enemy

Phagocytosis is a fundamental process used for nutrition in unicellular organisms and is present in various cell types in multicellular organisms. However, it is primarily a domain of specialized cells known as professional phagocytes, including macrophages, neutrophils, monocytes, dendritic cells, and osteoclasts (Levin et al, 2016). These cells excel at efficiently engulfing microorganisms and presenting antigens to lymphocytes, initiating adaptive immune responses (Nauseef&Borregaard, 2014).

In contrast, non-professional phagocytes like fibroblasts, epithelial cells, and endothelial cells can perform phagocytosis but with lower efficiency. They aren't as effective at ingesting microorganisms but play a very crucial role in eliminating dead cells and maintaining homeostasis (Canton, 2014).

Phagocytosis involves the sensing and ingestion of particles larger than 0.5 µm. The ingested particle is enclosed in a specialized organelle called the phagosome. Subsequently, the phagosome undergoes structural and content changes in a process known as phagosome maturation. Ultimately, the phagosome fuses with lysosomes to form a phagolysosome, containing enzymes capable of breaking down the ingested particle (Gordon, 2016).

The following set of images summarizes various aspects of phagocytosis:

3.12.2. Fc Receptor (FcR) Signaling

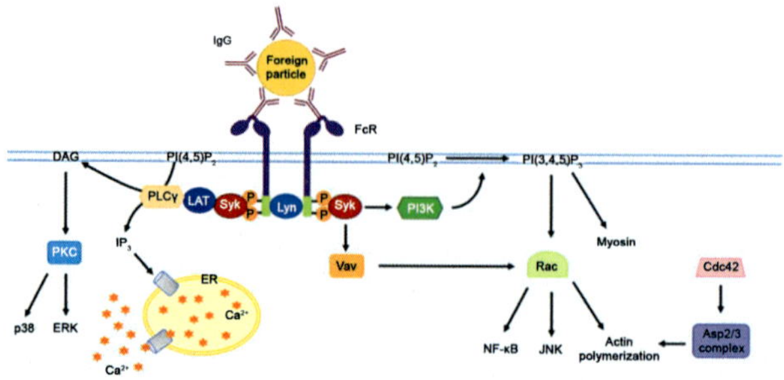

Figure 3.15. Various aspects of phagocytosis.

Source: Lee H-J, Woo Y, Hahn T-W, Jung YM, & Jung YJ, (2020). Formation and maturation of the phagosome: a key mechanism in innate immunity against intracellular bacterial infection. Microorganisms,8(9),1298. https://doi.org/10.3390/microorganisms8091298: © 2020 by the authors. Licensee MDPI, Basel, Switzerland. This article is an open access article distributed under the terms and conditions of the Creative Commons Attribution (CC BY) license (http://creativecommons.org/licenses/by/4.0/).

3.12.3. Complement Receptor (CR) Signaling

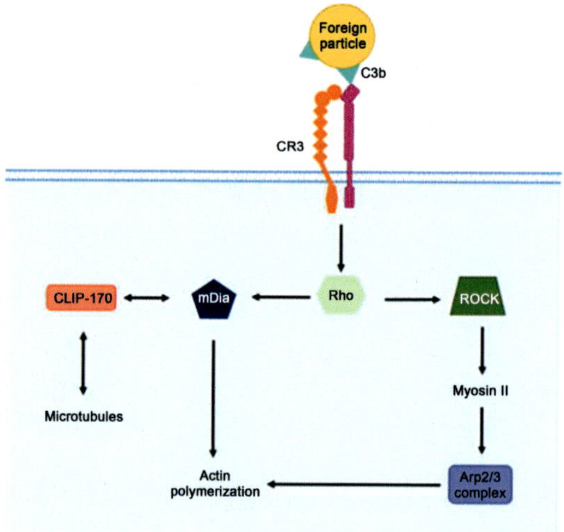

Figure 3.16. Different aspects of phagosomes

Source: Lee HJ, Woo Y, Hahn TW, Jung YM, Jung YJ, (2020). Formation and maturation of the phagosome: A key mechanism in innate immunity against intracel-

lular bacterial infection. Microorganisms, 8(9), 1298. https://doi.org/10.3390/microorganisms8091298: © 2020 by the authors. Licensee MDPI, Basel, Switzerland. This article is an open access article distributed under the terms and conditions of the Creative Commons Attribution (CC BY) license (http://creativecommons.org/licenses/by/4.0/).

3.12.4. Phagosome Maturation

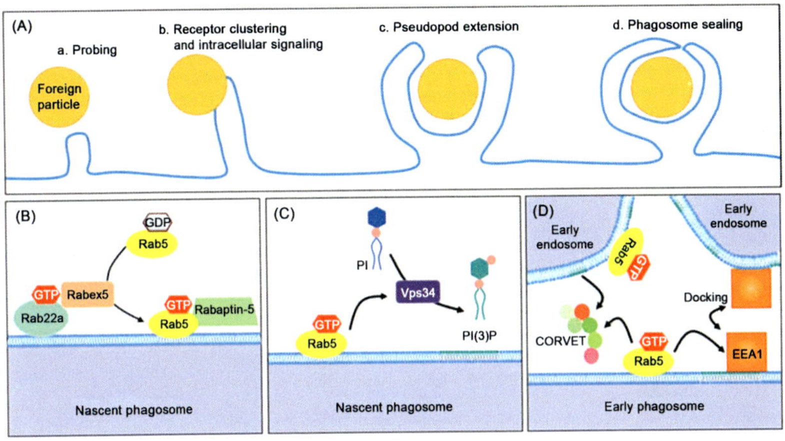

Figure 3.17. The process of phagosome maturation

Source:Lee HJ, Woo Y, Hahn TW, Jung YM, & Jung YJ, (2020). Formation and maturation of the phagosome: A key mechanism in innate immunity against intracellular bacterial infection. Microorganisms, 8(9), 1298. https://doi.org/10.3390/microorganisms8091298: © 2020 by the authors. Licensee MDPI, Basel, Switzerland. This article is an open access article distributed under the terms and conditions of the Creative Commons Attribution (CC BY) license (http://creativecommons.org/licenses/by/4.0/).

3.12.5. Formation of the Phagolysosome

Phagocytes can recognize various particles, including apoptotic cells and microbial pathogens, through specific receptors. These receptors can be non-opsonic or opsonic. Non-opsonic receptors directly detect molecular patterns on the target particle. Opsonic receptors identify host-derived proteins bound to the particle, such as antibodies, fibronectin, complement components, or lectins (Dambuzza, 2015).

A scheme of phagocytosis is given in Figure 3.20.

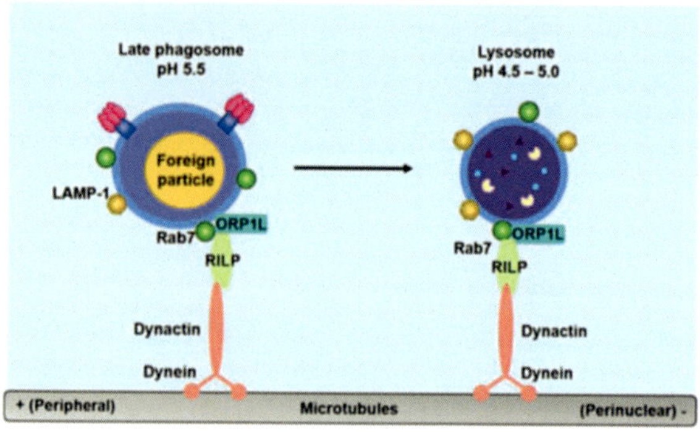

Figure 3.18. Formation of the phagolysosome

Source: Lee HJ, Woo Y, Hahn T-W, Jung YM, & Jung YJ, (2020). Formation and maturation of the phagosome: A key mechanism in innate immunity against intracellular bacterial infection. Microorganisms, 8(9):1298. https://doi.org/10.3390/microorganisms8091298: © 2020 by the authors. Licensee MDPI, Basel, Switzerland. This article is an open access article distributed under the terms and conditions of the Creative Commons Attribution (CC BY) license (http://creativecommons.org/licenses/by/4.0/).

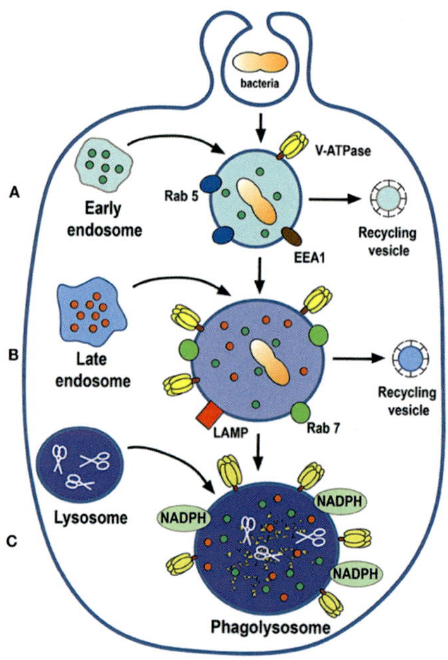

Figure 3.20.*A scheme of phagocytosis.*

Source: Uribe-Querol E & Rosales C, (2020). Phagocytosis: Our current understanding of a universal biological process. Front. Immunol., 11, 1066. doi: 10.3389/

Upon binding to the target, phagocytic receptors trigger signaling pathways that lead to changes in the actin cytoskeleton and membrane lipids. These changes cause the membrane to extend and cover the particle, ultimately leading to its internalization within the phagosome. Phagosome formation involves the sequential binding of phagocytic receptors and their coordinated action. The early phagosome then undergoes a series of fusion and fission events with endocytic vesicles to become a late phagosome. This late phagosome eventually fuses with lysosomes, transforming into a potent antimicrobial phagolysosome—a process known as phagosome maturation (Netea et al., 2016).

The initial step in the process of phagocytosis is the detection of the target particle. This detection is made possible through specialized receptors found on phagocytic cells. These receptors have the crucial role of identifying the target particle. There are two primary categories of receptors involved in this detection process:

- **Pattern-Recognition Receptors (PRRs):** These receptors directly recognize specific molecular patterns on the target particle known as pathogen-associated molecular patterns (PAMPs). Some PRRs have the unique ability to not only recognize PAMPs but also initiate the phagocytosis process. As a result, they are classified as non-opsonic receptors for phagocytosis. These receptors act as sensors for the presence of foreign invaders, and their activation triggers the engulfment of the target particle (Rosales & Uribe, 2013).
- **Opsonic Receptors:** In contrast to PRRs, opsonic receptors indirectly detect foreign particles. These receptors are specialized for binding to host-derived proteins, called opsonins, which have attached themselves to the target particle. Opsonins effectively "label" the target for phagocytosis. The two most well-known opsonic receptors are those for antibodies and complement components (Bakema& Van Egmond, 2011).
- **Antibody Receptors (FcRs):** These receptors specifically bind to the Fc (constant) region of antibodies, such as IgG and IgA. When antibodies have bound to a target particle, FcRs on phagocytic cells

recognize and engage with the antibody-coated particle, initiating phagocytosis.

- **Complement Receptors (CRs):** Complement is a cascade of proteins that can be activated by the presence of pathogens or foreign particles. Activated complement components, like iC3b, can bind to the surface of the target particle. Phagocytic cells express complement receptors (CRs), which recognize these activated complement components, marking the particle for engulfment.

A summary of receptors has been presented in Figure 3.21.

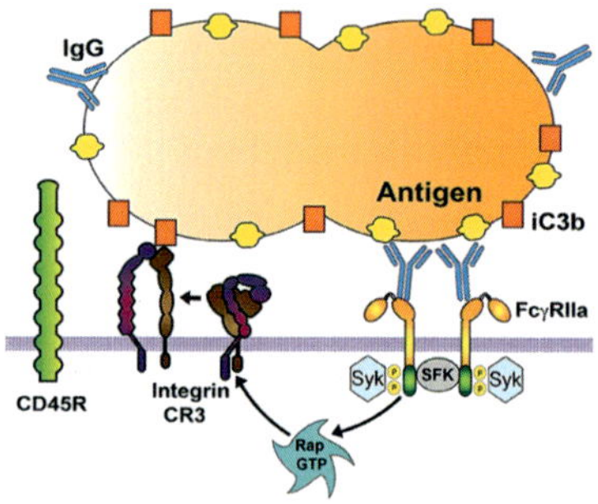

Figure 3.21. *A summary of receptors.*

Source: Uribe-Querol E & Rosales C, (2020) Phagocytosis: Our current understanding of a universal biological process. Front. Immunol., 11, 1066. doi: 10.3389/fimmu.2020.01066:

3.13. RELEASE OF ANTIMICROBIAL PROTEINS

Antimicrobial proteins and peptides (APPs) are a fascinating group of cationic molecules that play a crucial role in the body's defense against infections (Brown & Gordon, 2001). These molecules are primarily released by various immune cells, including neutrophils, monocytes, and macrophages. Additionally, they are produced by epithelial cells found in the skin and mucosal surfaces of the respiratory, gastrointestinal, and urinary tracts. This widespread production means that APPs are present in bodily fluids such as saliva, tears, nasal secretions, gastric juice, sweat, semen, airway surface liquid, and breast milk (Azad et al., 2008).

A few examples of AMPs have been presented in Table 3.2.

Table 3.2. Examples of AMPs

Category A	Peptides	Unique Structural/ Sequence Feature	Source
α helical peptides	Aurein 1–2 [36, 37]	Amidated C-terminus	Frogs
	Mellitin (Ericson, Duffau, Yasuda, Ortiz-Lopez, Rothamel, Rifkin, et al. (2014)	Amidated C-terminus	Bees
	Brevinin 1 [48]	–	Frogs
	Maculatins [50]	Amidated C-terminus	Frogs
	Citropin [51]	Amidated C-terminus	Frogs
	Buforin II [52]	–	Toad
	Cathelicidins [31]	–	–
	LL-37b	Amidated C-terminus	Humans
	BMAP [27, 28, 34][b]	–	Bovine
	Magainins	–	Frogs
	Cecropin	Amidated C-terminus	Insect
β sheet peptides	Cathelicidins [31]	–	–
	Protegrins	Cysteine rich	Pigs
	Bactenecin	Disulfide forming loop/Arginine rich	Bovine
	Defensins c (Zhang, Schwartz, Pantelic, Li, Knazze, Nobile, et al. (2006)) [53, 54, 55]	–	–
	α defensins	Three disulfide bonds	Mammals
	β defensins	Three disulfide bonds	Mammals

	θ defensins	Three disulfide bonds and cyclic	Gorilla
	Tachyplesins (Gordon, (2016)) and Polyphemusin [49]	Cysteine/arginine rich and amidated C-terminus	Horse Crab
Extended/ flexible	Cathelicidins [31]	–	–
	PR-39[b]	Proline and arginine rich	Pigs
	Tritrpticin	Tryptophan and arginine rich	Pigs
	Indolicidin	Tryptophan and amidated C-terminus	Bovine
	Crotalicidin [15–34]	Lysine rich	Snakes
	Histatins [56]	Histidine rich and amidated C-terminus	Humans

Source: Kumar P, Kizhakkedathu JN, Straus SK, (2018). Antimicrobial peptides: Diversity, mechanism of action and strategies to improve the activity and biocompatibility in vivo. Biomolecules, 8(1), 4. https://doi.org/10.3390/biom8010004:

3.14. SOME CLINICALLY IMPORTANT APPS IN EARLY LIFE

- **Defensins:** These are cationic peptides found in various organisms, including humans. Human defensins include α-defensins (HNP-1, HNP-2, HNP-3, HNP-4, HD-5, and HD-6) and β-defensins (HBD-1, HBD-2, and HBD-3).
- **Cathelicidins:** These are multifunctional bactericidal peptides. One of the well-known cathelicidins in humans is LL-37.
- **Protegrins:** These are porcine APPs with structural similarities to cathelicidins. They have served as templates for the development of therapeutics for use in humans.
- **Bactericidal/Permeability-Increasing Protein (BPI):** BPI is an anti-infective protein stored within the primary granules of human

polymorphic neutrophils (PMNs). It has been developed as a synthetic therapeutic.

- **S100 Proteins:** Calprotectin is a predominantly neutrophil-derived metal-chelating protein belonging to the S100 protein family.
- **Lactoferrin (LF):** LF is a protein found in neutrophils and mammalian milk, and it plays a role in iron binding and immune defense.
- **Lysozyme:** Lysozyme is an enzyme with antimicrobial properties that can break down bacterial cell walls.
- **RNAses:** These include RNases 4, 5, and 7.

APPs can be constitutively expressed, meaning they are continuously produced, or inducible, meaning their production can be triggered in response to proinflammatory stimuli. These molecules have multiple functions in the immune response, including:

3.15. DIRECT ANTIMICROBIAL ACTION AGAINST INVADING PATHOGENS

- Immunomodulatory functions.
- Inducing angiogenesis and promoting wound healing.
- Inhibiting proinflammatory responses triggered by bacterial products.
- Modulating adaptive cellular immune responses.
- Mediating immune cell development in various tissues.
- Acting as chemo-attractants for other immune cells.

In terms of their antimicrobial activity, APPs target invading bacteria by initially making electrostatic contact with the anionic surface of the bacterial cell. The specific mechanisms of action can vary among different APP families, but a common critical step in their antimicrobial activity involves the permeabilization of the target bacteria's cytoplasmic membranes (Netea et al., 2016).

Recent research has also highlighted the role of APPs in the extracellular entrapment of bacteria. Neutrophil extracellular traps (NETs), composed of extracellular DNA, chromatin, and APPs, are described as lattices that mediate the extracellular killing of bacteria. Similar processes occur at mucosal surfaces, where defensins form nanonets to trap bacteria, preventing their invasion into deeper tissues.

Many APPs have shown potential as standalone therapeutics or adjunctive agents to reduce the duration of antibiotic treatment or inflammation induced by killed microbes or their products. Some APPs have undergone clinical trials, demonstrating their potential as therapeutic agents to combat infections.

Thus, antimicrobial proteins and peptides are essential components of the body's defense against infections, with diverse functions that extend beyond direct antimicrobial action. Their wide-ranging presence in various bodily fluids and tissues highlights their importance in maintaining health and combating microbial threats (Zhang et al., 2009).

3.16. NEUTROPHIL EXTRACELLULAR TRAPS (NETS)

Neutrophil Extracellular Traps (NETs) are fascinating structures formed by neutrophils as a part of the immune response to various challenges, and they play several critical roles in both host defense and disease development. NETs can be released through various pathways, including vesicle-mediated release, cell lytic release, and mitochondrial release. They are released with the help of three distinct mechanisms, noted herewith:

3.17. VESICLE-MEDIATED RELEASE OF NEUTROPHIL EXTRACELLULAR TRAPS

In response to external stimulation, such as exposure to pathogens like *Staphylococcus aureus*, neutrophils undergo specific changes. The nucleus, which typically has a lobular shape in neutrophils, becomes round. Chromatin, which is the complex of DNA and associated proteins, is uniformly concentrated within the nucleus. The nuclear membrane is then ruptured, and vesicles containing DNA are formed. These vesicles are wrapped with DNA. These DNA-containing vesicles move towards the cell membrane and subsequently fuse with it, releasing the DNA into the extracellular space. This release of DNA forms NETs. Vesicle-mediated NET release is relatively rapid, occurring within 30 to 60 minutes after stimulation (Busso& So, 2012) (Figure 3.21).

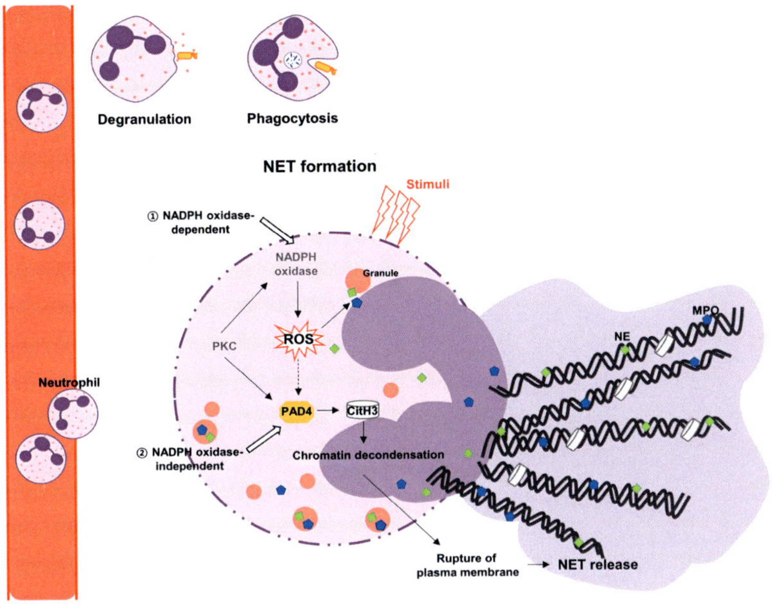

Figure 3.21. *A scheme of NET formation and functioning.*

Source: Jo A, & Kim DW, (2023). Neutrophil extracellular traps in airway diseases: Pathological roles and therapeutic implications. International Journal of Molecular Sciences, 24(5), 5034. https://doi.org/10.3390/ijms24055034: © 2023 by the authors. Licensee MDPI, Basel, Switzerland. This article is an open access article distributed under the terms and conditions of the Creative Commons Attribution (CC BY) license (https://creativecommons.org/licenses/by/4.0/).

3.18. CELL LYTIC NET FORMATION

In certain situations, NETs can be formed through a process that leads to cell lysis, or the breaking apart of the neutrophil. Prototypical formation of NETs in response to a potent stimulant like phorbol myristate acetate (PMA) can lead to cell lysis. This process requires more time, typically around 3 to 4 hours. During cell lytic NET formation, the nuclear membrane is enzymatically degraded, and this degradation process is triggered by external stimulation. Neutrophil elastase (NE) and myeloperoxidase (MPO) are enzymes that play a role in nuclear lysis and chromatin decondensation during PMA-induced NET formation (Netea et al., 2016).

3.19. MITOCHONDRIAL DNA RELEASE

Some studies have suggested that neutrophils can release mitochondrial DNA to form NETs in a reactive oxygen species (ROS)-dependent manner. The detailed molecular mechanisms of mitochondrial DNA release in NET formation are still not fully understood. This pathway is less well-characterized compared to the other two pathways, and further research is needed to clarify the mechanisms involved (Chen et al, 2014).

It is important to note that the formation of NETs is a dynamic and complex process that can vary depending on the specific stimuli and conditions. The ability of neutrophils to release NETs through different pathways adds to their versatility in responding to various types of infections and threats. Understanding these pathways is crucial for studying the role of NETs in host defense and various diseases (Clerk et al, 2007).

Various triggers for NET formation have been summarized Table 3.3 by a group of researchers Klopf et al (2021) in the *International Journal of Molecular Sciences.*

Table 3.3. Triggers for NET formation

Category	Disease	Identified Triggers
Cardiovascular	Atherosclerosis	IL-8, IL-1β, crystalized cholesterol, oxLDL
	Thrombosis	HMGB-1, pathophysiologic hemodynamic forces
	Abdominal aortic aneurysm	IL-1β, *Porphyromonasgingivalis*
Autoimmune & Autoinflammatory	Systemic lupus erythematosus & lupus nephritis	IFN-α, ANCAs, RNPs, acetylated histones & apoptotic microparticles
	Psoriasis	IL-8, IL-17, IL-23, TNF-α, autoantigens (keratin 17, LL37)
	Rheumatoid arthritis & arthritis urica	ACPAs, ANCAs, RNPs, IL-17A, TNF-α, hyperuricemia and monosodium urate crystals
	Small-vessel vasculitis	ANCAs
	Diabetes mellitus type I	Hyperglycemia, β-cell death & autoantibodies
	Crohn's disease & ulcerative colitis	enhanced ROS production [34,35]

Metabolic	Obesity & diabetes mellitus type II	Hyperglycemia, IL-6, IL-8, TNF-α, homocysteine
Malignant neoplasia	metastatic breast cancer cells, metastatic colorectal cancer	G-CSF, IL-8, intratumoral hypoxia
Infectious	Sepsis	HMGB-1, eCIRP, oxLDL
	COVID-19	HMGB-1, hypoxia

Key: IL: interleukin; HMGB-1: high-mobility group box 1 protein; IFN-α: interferon α; ANCA: anti-neutrophil cytoplasmic antibodies; RNP: anti-ribonucleoprotein (antibodies); TNF-α: tumor necrosis factor-α; ACPA: anti-citrullinated protein antibodies; ROS: reactive oxygen species; G-CSF: granulocyte colony-stimulating factor; eCIRP: extracellular cold-inducible RNA-binding protein; oxLDL: oxidized low-density lipoprotein.

Source: Klopf J, Brostjan C, Eilenberg W, &Neumayer C, (2021). Neutrophil extracellular traps and their implications in cardiovascular and inflammatory disease. International Journal of MolecularSciences, 22(2), 559. https://doi.org/10.3390/ijms22020559: © 2021 by the authors. Licensee MDPI, Basel, Switzerland. This article is an open access article distributed under the terms and conditions of the Creative Commons Attribution (CC BY) license (http://creativecommons.org/licenses/by/4.0/).

3.20. STRUCTURAL COMPONENTS OF NETS

Similarly, it is interesting to note certain structural components of NETs, as mentioned herein (Clerk et al., 2007):

- **DNA:** DNA is the primary structural component of NETs. The DNA filaments in NETs have a diameter of approximately 15–17 nanometers.
- **Proteins:** NETs contain a variety of proteins, including:
- **Histones:** Histones H1, H2A, H2B, H3, and H4 are found in NETs. Histones are structural proteins associated with DNA in the cell nucleus.
- **Neutrophil Elastase (NE):** NE is an enzyme released by neutrophils and is involved in the breakdown of proteins. It binds to NETs.
- **Myeloperoxidase (MPO):** MPO is an enzyme that plays a role in antimicrobial activity and is found on NETs.
- **Cathepsin G:** This is another enzyme present on NETs.
- **Neutrophil Defensins:** These are antimicrobial peptides produced by neutrophils.

- **Calprotectin:** Also known as the S100A8/A9 complex, calprotectin is a cytoplasmic protein found in NETs.

The core proteins of NETs are histones H2A, H2B, H3, and H4, which make up about 70% of the total NET proteins. NE, S100A8, lactotransferrin, azurocidin, and cathepsin G are among the other proteins found in NETs. Additionally, matrix metalloproteinase-9 (MMP-9), tissue factor (TF), von Willebrand factor (vWF), and fibrinogen have also been identified in NETs.

3.21. DETECTION OF NETS

There are two main methods for identifying NETs:

- **DNA Staining with SYTOX Green:** This method involves staining the secreted DNA with a nucleic acid dye called SYTOX Green. The stained DNA can then be observed under a fluorescence microscope, and the fluorescence intensity can be measured using a microplate reader. This approach is relatively simple and direct but is primarily used under specific circumstances.
- **Immunofluorescence Detection:** This method involves detecting specific components of NETs, including DNA, citrullinated histone H3 (citH3), and myeloperoxidase (MPO) or neutrophil elastase (NE), using immunofluorescence techniques. In cells or tissues, the detection of NETs typically involves a combination of DNA staining, citH3, and MPO or NE staining.

  These detection methods allow researchers to visualize and quantify NETs in various biological samples. They are valuable tools for studying the role of NETs in immune responses and various diseases.

3.22. FUNCTIONAL CONTRIBUTION OF NETS

They play several critical roles in both host defense and disease development. Here's an overview of the multifaceted roles of NETs:

- **Host Defense:** NETs are primarily a host defense mechanism. Neutrophils, a type of white blood cell, release NETs to capture and neutralize pathogens like bacteria, fungi, and parasites. These structures consist of DNA strands studded with antimicrobial proteins and granular contents, creating a physical barrier to entrap and kill microbes.
- **Sepsis:** In sepsis, which is a severe and often life-threatening response to infection, neutrophils can accumulate at the site of infection and release NETs. While this may initially serve as a

defense mechanism, excessive NET formation can contribute to the development and progression of sepsis (Erison et al., 2014).

- **Coagulation and Thrombosis:** NETs can act as a scaffold for the binding of platelets, red blood cells, and plasma proteins. This interaction can activate both cellular and plasmatic coagulation systems, potentially leading to thrombosis, which is the formation of blood clots within blood vessels (Fuchs et al., 2010).
- **Atherosclerosis:** NETs have been implicated in the formation and development of atherosclerotic plaques. These plaques are a hallmark of cardiovascular disease. Studies have shown that treatment with deoxyribonuclease I (DNase I), an enzyme that breaks down DNA, can significantly reduce plaque size in experimental models (Garg et al., 2015a).
- **Autoimmune Diseases:** NETs may play a role in the pathogenesis of autoimmune diseases. They can induce the body to generate autoantibodies against autoantigens, contributing to conditions where the immune system mistakenly attacks the body's own tissues. This includes diseases like lupus and rheumatoid arthritis (Garg et al., 2015b). Granger et al. (2019) reviewed the involvement of NETs in autoimmunity and has been depicted in Figure 3.22.

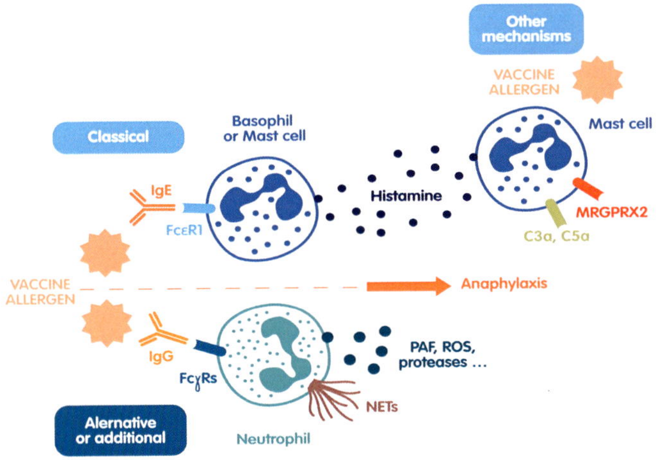

Figure 3.22. *The involvement of NETs in autoimmunity.*

Source: Granger V, Peyneau M, Chollet-Martin S & de Chaisemartin L, (2019) Neutrophil extracellular traps in autoimmunity and allergy: Immune complexes at work. Front. Immunol. 10:2824. doi: 10.3389/fimmu.2019.02824: Copyright © 2019 Granger, Peyneau, Chollet-Martin and de Chaisemartin. This is an open-

3.23. DIABETES, ALZHEIMER'S DISEASE, AND TUMOR PROGRESSION

NETs have also been implicated in other conditions such as diabetes, Alzheimer's disease, and tumor progression. Their role in these diseases is complex and not fully understood, but they appear to contribute to inflammation and tissue damage (Garg et al., 2015b). Shafqat et al. (2023) reviewed the involvement of NETs in neurological conditions as shown in Table 3.4.

Table 3.4. The Involvement of NETs in Neurological Conditions

Infections	CSF analysis of patients with acute pneumococcal meningitis, Lyme neuroborreliosis, and viral meningitis reveals the presence of NETs (58, 154). DNase-1 treatment significantly reduces bacterial load in the brain, liver, spleen, and blood of rat models of pneumococcal meningitis from the SP001 strain isolated from patients (58).
Trauma	Leukocyte adhesion molecules and chemoattractants in contusional and peri-contusional brain tissue facilitate neutrophil adhesion and extravasation, releasing NETs. Cl-amidine and DNase-1 reduced CNS NETs, cerebral edema, improved cerebral blood flow, and improved neurologic function post-TBI (70).
	Paroxysmal sympathetic hyperactivity (PSH) increases morbidity and mortality risk in TBI patients and correlates with NETs in the paraventricular nucleus. The protein LL-37 in NETs activates microglial cells to release IL-1β, which may mediate PSH (73).
	Neutrophil infiltration and NET production occur after spinal cord injury (SCI) in rat models, associated with neuroinflammation and spinal cord edema, and prevented by Cl-amidine and DNase-1. NETs impair blood-spinal cord barrier integrity by degrading tight junctions and upregulating TRPV4 expression on endothelial cells. NET inhibition by Cl-amidine or DNase-1 promote functional motor recovery post-SCI (155).
Stroke	In mice, CNS neutrophil recruitment to peri-infarct areas and PAD-4-dependent NET production *via* PAD-4 peaks 3–5 days after stroke. DNase-1 or Cl-amidine/PAD-4 deficiency reduce BBB permeability and increase vascular remodeling. Levels of type I IFN increased 10-fold in the ischemic cortex and were decreased by Cl-amidine, indicating NET-induced type I IFN production, increasing BBB permeability (54).
	Brain specimens from patients who died from ischemic stroke reveal dense neutrophilic infiltration and NET production in the ipsilateral brain tissue. NETs are detected in platelet-rich areas in ischemic stroke thrombi. HMGB1 expression is increased on the surface of platelets and in the plasma of stroke patients, induces NET production, and exacerbates ischemic brain injury (78).
Alzheimer's disease	Neutrophils adhere to BMECs and extravasate into the brain of 9–13 months-old 5xFAD mice, which show Aβ accumulation in the brain. Intraparenchymal neutrophils specifically migrated toward areas of Aβ plaques (120).
	Neutrophil adhesion, extravasation, and NETs were seen in the brain areas of Aβ deposits in 5xFAD mice. Soluble Aβ induced LFA-1 integrin expression on neutrophils (causing adhesion to BMECs) and activated ROS production. Depleting neutrophils by anti-Ly6G or inhibiting LFA-1 significantly improved cognitive performance, reduced microgliosis, and lowered the load of Aβ and phosphorylated tau in 3xTg-AD mice (121).
Glioma	Immunofluorescence assays for MPO and cit-H3 demonstrate higher presence of NETs in grade IV gliomas than in grade II and III tumors. NETs activate RAGE receptors on glioma cells to enhance proliferation, migration, invasiveness, and IL-8 production *in vitro*. Tumor-derived IL-8 drives neutrophil recruitment to the tumor and NET production (138).

Source: Shafqat A, Noor Eddin A, Adi G, Al-Rimawi M, Abdul Rab S, Abu-Shaar M, Adi K, Alkattan K, & Yaqinuddin A, (2023). Neutrophil extracellular traps in central nervous system pathologies: A mini review. Front. Med., 10, 1083242. doi: 10.3389/fmed.2023.1083242:

- **Pancreatitis:** In pancreatitis, a condition characterized by inflammation of the pancreas, NETs may occlude glandular ducts, further exacerbating the disease (Garg et al., 2015b).
- **Cytokine Release:** Cytokine release is a fundamental process in the immune response and plays a pivotal role in regulating inflammation, immune cell communication, and various physiological and pathological processes. Cytokines are signaling molecules produced by a wide range of cells, including immune cells, to communicate with each other and orchestrate the immune response. Here, we'll explain cytokine release in detail.

3.24. CYTOKINES OVERVIEW

Cytokines are a diverse group of small proteins or glycoproteins that act as messengers in the immune system and other tissues. They are produced by immune cells (such as T cells, B cells, macrophages, and dendritic cells), as well as non-immune cells like fibroblasts and endothelial cells. Cytokines can have autocrine (affecting the cell that produces them), paracrine (affecting nearby cells), or endocrine (affecting distant cells) effects.

3.25. CYTOKINE FUNCTIONS

- **Inflammation:** Cytokines play a central role in initiating and regulating the inflammatory response. They promote the recruitment of immune cells to sites of infection or tissue damage (Giepmans et al., 2006).
- **Immune Cell Activation:** Cytokines activate various immune cell types. For example, interferons activate macrophages and enhance natural killer cell activity (Lippincott-Schwartz et al., 2004).
- **Immune Cell Differentiation:** Some cytokines, like interleukins, drive the differentiation of immune cells into specific subsets, such as T helper cells, cytotoxic T cells, and B cells.
- **Cell Proliferation and Differentiation:** Cytokines regulate the proliferation and differentiation of immune and non-immune cells. For instance, interleukin-7 stimulates the production of B cells from hematopoietic stem cells.
- **Antiviral and Antibacterial Defense:** Interferons induce an

antiviral state in host cells and promote the destruction of infected cells.

- **Tissue Repair and Growth:** Certain cytokines, such as transforming growth factor-beta (TGF-β), are involved in tissue repair and growth.
- **Regulation of the Immune Response:** Cytokines help maintain a balance between pro-inflammatory and anti-inflammatory responses, preventing excessive inflammation.

3.26. CYTOKINE RELEASE PROCESS

Cytokines are typically synthesized in response to immune stimuli. For example, when immune cells recognize pathogens (e.g., bacteria, viruses) or inflammatory signals, they start producing cytokines. The synthesis of cytokines involves the activation of genes encoding specific cytokines and their subsequent translation into protein molecules (Nauseef&Borregaard, 2014). Once synthesized, cytokines are stored intracellularly or transported to the cell surface. Upon activation (often by immune cell receptor signaling), cytokines are released into the extracellular environment. This release can occur through various mechanisms, including exocytosis (for pre-formed cytokines), membrane shedding, or active secretion. Released cytokines then bind to specific receptors on target cells, initiating signaling pathways that trigger various cellular responses (Pushparaj et al., 2009).

3.27. REGULATION AND FEEDBACK

Cytokine production and release are tightly regulated to prevent excessive inflammation and immune responses. Negative feedback mechanisms exist to limit cytokine production once the immune response has achieved its goal. Dysregulation of cytokine release can lead to chronic inflammation, autoimmune diseases, or cytokine storm, a severe and often life-threatening immune response seen in conditions like sepsis (Reefman et al., 2010).

Conclusively, cytokine release is a critical component of the immune response and serves as a means of communication between immune cells and other cells in the body. The balance and regulation of cytokine production and release are essential for effective immune responses and maintaining overall health.

In order to show the complexity of cytokines in health, two examples are presented Table 3.5. The first is a brief summary of cytokines that play a role in bone disease as compiled by Kany et al. (2019).

Table 3.5. The roles of cytokines in health-as seen in bone disease

Species	Study	Cell Type/Organ	Major Finding
Human	*In vivo*	bone	IL-6 is a predictor of postmenopausal bone loss
	Meta-analysis	bone	GG genotype of IL-6–634C/G polymorphism seems to play a role in reducing bone mineral density
	Meta-analysis	bone	IL-6–634C/G and IL-6–174G/C polymorphisms lead to modest effects on bone mineral density
	Meta-analysis	bone	CC genotype of IL-6 G-174C polymorphism may be associated with high bone mineral density at femoral neck and distal radius and decreased risk of osteoporosis in the Caucasian population
	In vivo	bone	IL-6 G-174C promoter polymorphism may be a genetic marker for bone loss and wrist fracture among older women
	Meta-analysis	bone	IL6–174 G/C gene polymorphism positively correlated with osteoporosis risk
	In vivo	bone	Variation within the low levels of IL-6 predicts bone loss and resorption
	In vitro	whole blood cells	Increased IL-6 production by whole blood cells from postmenopausal women with osteoporosis compared to controls

	In vivo	bone	IL-6 is upregulated in postmenopausal women with low bone mineral density compared to postmenopausal women with normal bone mineral density
	Chronic obstructive pulmonary disease (COPD) patients	bone	RANKL (Receptor activator of NF-κB ligand)-expressing neutrophils correlate negatively with bone marrow density. Plasma levels of IL-6 are increased in COPD patients and correlate with RANKL expression by neutrophils
Mouse	*In vivo*	osteoclasts	IL-6 mediates stimulation of osteoclastogenesis after estrogen loss
Human	*In vitro*	whole blood cells	Increased IL-1 beta production by whole blood cells from postmenopausal women with osteoporosis compared to controls
	COPD patients	bone	RANKL-expressing neutrophils correlate negatively with bone marrow density. Plasma levels of IL-1 beta are increased in COPD patients and correlate with RANKL expression by neutrophils
	In vivo	bone	IL-1β (-511C/T) polymorphism is associated with pathogenesis of osteoporosis in postmenopausal women
	In vivo	bone	Serum IL-1β significantly higher in women with osteoporosis than controls

	In vivo	bone	Serum IL-1 is significantly reduced after treatment of postmenopausal with calcitriol
	In vitro	osteoblasts	IL-1beta and Tumor necrosis factor-alpha (TNF-α) regulate osteoblast cell number by up-regulating the Fas-mediated apoptosis of osteoblasts
Rat	*In vivo*	bone	Interleukin-1 receptor antagonist decreases bone loss and bone resorption in a rat model of postmenopausal osteoporosis
Human	*In vivo*	bone	IL-33 levels in postmenopausal women significantly lower compared to healthy controls, positively correlated with serum parathyroid hormone and inverse correlated with C-terminal telopeptide of type 1 collagen
Human	*In vitro*	whole blood cells	Increased TNF-α production by whole blood cells from postmenopausal women with osteoporosis compared to controls
	In vivo	bone	TNF-α is upregulated in postmenopausal women with low bone mineral density compared to postmenopausal women with normal bone mineral density
	In vivo + in vitro	osteoclasts	Estrogen deficiency $\rightarrow$ TNF-α and RANKL $\uparrow$ $\rightarrow$ osteoclast formation and number of osteoclast precursors $\uparrow$

	In vivo	bone	Association between TNF-α-308G>A polymorphism and postmenopausal osteoporosis
	In vivo	bone	Serum TNF-α is significantly reduced after treatment of postmenopausal with calcitriol
	In vivo; in vitro		TNF-α increased in postmenopausal women with osteoporosis and highly correlated with the RANK and estrogen levels TNF-α synergistically promotes RANKL-induced osteoclasts formation through activation of Phosphoinositide 3-kinases (PI3K)/Akt signaling
	In vitro	mesenchymal stem cells (MSC)	TNF-α suppresses osteogenic differentiation of MSCs by accelerating P2Y2 receptor in estrogen-deficiency induced osteoporosis
	In vitro	osteoblasts	IL-1beta and TNF-α regulate osteoblast cell number by up-regulating the Fas-mediated apoptosis of osteoblasts
Mouse	*In vivo*	bone marrow-derived mesenchymal stem cells (BMMSCs)	TNF-α inhibits Forkhead box protein O1 (FoxO1) and thereby aggravates oxidative damage in BMMSCs during osteoporosis
Rat	*In vitro*	bone cultures of fetal rat calvariae	TNF-α causes osteoclastic bone resorption and inhibits bone collagen synthesis

Human	patients with systemic lupus erythematosus	bone	IL-10 level is elevated in systemic lupus erythematosus patients with osteoporosis
	In vivo	bone	IL-10 gene-597 C>A polymorphism is associated with higher risk for osteoporosis
	In vivo	bone	Lower levels of IL-10 in postmenopausal women with osteoporosis
	In vivo	bone	Association between IL10–1082G>A polymorphism and postmenopausal osteoporosis
Mouse	*In vitro*	RAW264.7 monocytes	IL-10 directly inhibits osteoclastogenesis is by suppressing nuclear factor of activated T-cells, cytoplasmic 1 activity
	In vivo	bone	IL-10 is important for promoting osteoblast maturation and reducing bone loss during early stages of type-1 diabetes
	In vivo	bone	IL-10−/− mice develop the reduced bone mass, increased mechanical fragility, and suppressed bone formation (hallmarks of osteoporosis)
Human	*In vivo*	bone	IL-8 significantly increase in post-menopausal women with osteoporosis and bone loss
	COPD patients	bone	RANKL-expressing neutrophils correlate negatively with bone marrow density, plasma levels of IL-8 are increased in COPD patients and correlate with RANKL expression by neutrophils

	In vitro	osteoblasts	IL-8 may contribute to osteoporosis in rheumatoid arthritis by enhanced osteoblast-mediated osteoclastogenesis (partly via IL-6 production)
Rats	*In vivo*	bone	IL-8 and bone loss reduction after atorvastatin treatment of rats with glucocorticoid-induced osteoporosis

Source: Kany S, Vollrath JT, Relja B, (2019). Cytokines in inflammatory disease. InternationalJournal of Molecular Sciences, 20(23), 6008. https://doi.org/10.3390/ijms20236008:

The second example is the involvement of cytokines in brain health and disease, as compiled by Boraschi et al (2023).

Table 3.6. Cytokines in health-brain disease

Cytokine/ receptor	Physiological role	Role in pathogenesis	Role in pathology
IL-1β	• Learning and memory processes • Thermoregulation	• Induction of inflammation • Induction of fever	• Persistence of inflammation • Neurodegeneration
IL-1Ra	• Unknown (regulation of IL-1 activity)	• Unknown	• Increased with disease (feedback mechanism)?
IL-18	• Neuroprotective	• Induction of inflammation	• Persistence of inflammation
IL-18BP	• Unknown (regulation of IL-18 activity)	• Unknown	• Increased with disease (feedback mechanism)?
IL-33	• Regulates microglia proliferation • Regulates the microglial inflammatory/anti-inflammatory balance	• Anti-inflammatory protective effects	• Upregulated in pathology • Negative correlation with disease severity
IL-37	• Unknown	• Anti-inflammatory protective effects	• Downregulated in active disease • Negative correlation with disease severity
IL-1R3b	• Accessory receptor mediating the neuroprotective effects of IL-1	• Unknown	• Unknown
IL-1R4	• Mediates the homeostatic activities of IL-33	• Unknown	• Soluble form upregulated during disease (feedback mechanism)?
IL-1R5	• Mediates the homeostatic activities of IL-18	• Associated with disease pathogenesis in EAE	• Unknown
IL-1R8	• Anti-inflammatory orphan receptor; physiological role unknown	• Unknown	• Necessary for the neuroprotective role of IL-37
IL-1R9	• Orphan receptor, involved in learning and memory processes	• Unknown (mutations are associated with X-linked mental retardation, schizophrenia and autism)	• Unknown

Source: Boraschi D, Italiani P, Migliorini P and Bossù P, (2023). Cause or consequence? The role of IL-1 family cytokines and receptors in neuroinflammatory and neurodegenerative diseases. Front. Immunol., 14, 1128190. doi: 10.3389/fimmu.2023.1128190.

3.28. ANTIBODY PRODUCTION, RECOGNITION, AND PRESENTATION

Antibody production, also known as humoral immunity, is a fundamental component of the adaptive immune response. It involves the generation of specific proteins called antibodies or immunoglobulins that target and neutralize pathogens such as bacteria, viruses, and toxins. Here's a detailed explanation of the antibody production process (Woodfin et al., 2011).

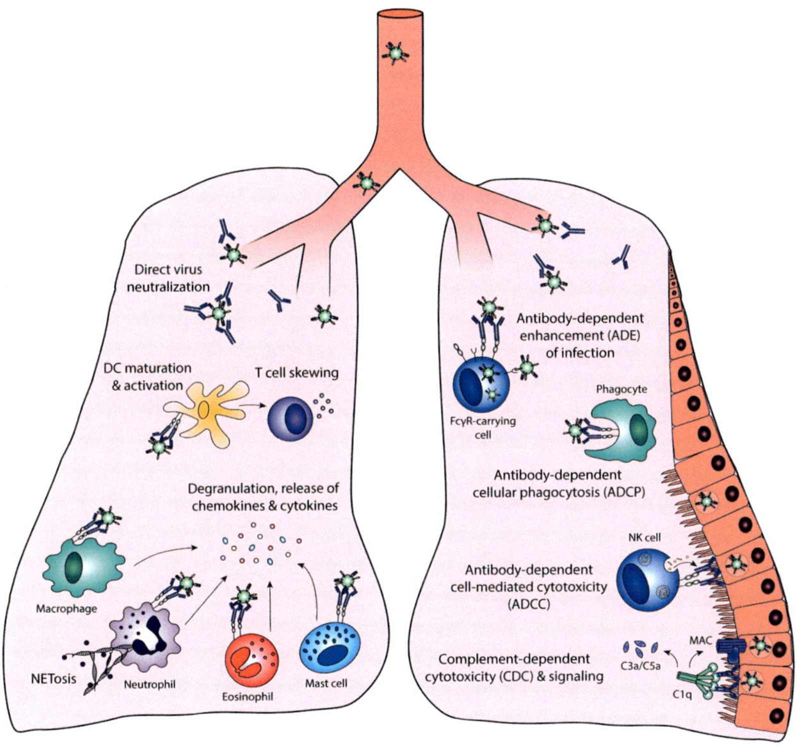

Figure 3.23. *The functioning of antibodies.*

Source: van Erp EA, Luytjes W, Ferwerda G and van Kasteren PB, (2019) Fc-mediated antibody effector functions during respiratory syncytial virus infection and disease. Front. Immunol. 10:548. doi: 10.3389/fimmu.2019.00548: Copyright

Antibody production begins when the immune system recognizes a foreign molecule called an antigen. Antigens are typically found on the surface of pathogens or foreign substances. Antigens can be proteins, carbohydrates, lipids, or other molecules that are not normally present in the body.

Antigen-presenting cells (APCs), such as dendritic cells, macrophages, and B cells, capture and process antigens. They break down the antigens into smaller fragments. APCs then present these antigen fragments on their cell surfaces using specialized proteins called major histocompatibility complex (MHC) molecules (Nauseef&Borregaard, 2014) (Figure 3.23).

3.28.1. Cell and B Cell Activation

Helper T cells (Th cells) are a type of immune cell that plays a crucial role in antibody production. When a helper T cell encounters an APC presenting an antigen that matches its T cell receptor, it becomes activated. Activated helper T cells release cytokines, which stimulate B cells.

B cells are specialized immune cells responsible for antibody production. Each B cell has a unique antigen receptor on its surface called a B cell receptor (BCR). When an antigen matches the BCR of a specific B cell, the B cell is activated. Helper T cells provide the necessary co-stimulatory signals to fully activate the B cell. Activated B cells undergo clonal expansion, dividing and producing a large number of identical B cells.

3.28.2. Antibody Class Switching

B cells initially produce antibodies of the IgM class. However, under the influence of cytokines from helper T cells, they can undergo class switching to produce other antibody classes like IgG, IgA, or IgE. This process allows antibodies to perform various functions in the immune response. In a research article by Horns et al. (2016), class switch recombination events documented in humans were summarized. They have been presented in Table 3.7.

Table 3.7. Antibody class switch recombination events documented in humans

	Source Class						
Destination Class	IgM/IgD	IgG3	IgG1	IgA1	IgG2	IgG4	IgE
IgG3	Known (Fujieda et al., 1995; Malisan et al., 1996)						
IgG1	Known (Fujieda et al., 1995; Malisan et al., 1996)	Novel					
IgA1	Known (Jabara et al., 1993; Zan et al., 1998)	Known (Lin et al., 2014)	Known (Zan et al., 1998)				
IgG2	Known (Malisan et al., 1996)	Novel	Novel	Novel			
IgG4	Known (Fujieda et al., 1995; Jabara et al., 1993)	Novel	Novel	Novel	Novel		
IgE	Known (Jabara et al., 1993; Xiong et al., 2012)	Not detected	Known (Xiong et al., 2012)	Novel	Not detected	Known (Jabara et al., 1993)	
IgA2	Known (He et al., 2007; Lin et al., 2014)	Known (Lin et al., 2014)	Known (Lin et al., 2014)	Known (He et al., 2007; Lin et al., 2014)	Known (Lin et al., 2014)	Novel	Not detected

Source: Felix Horns, Christopher Vollmers, Derek Croote, Sally F Mackey, Gary E Swan, Cornelia L Dekker, Mark M Davis, & Stephen R Quake, (2016). Lineage tracing of human B cells reveals the in vivo landscape of human antibody class switching eLife 5:e16578: © 2016, Horns et al: This article is distributed under the terms of the Creative Commons Attribution License, which permits unrestricted use and redistribution provided that the original author and source are credited.

3.28.3. Plasma Cell Differentiation

Activated B cells can differentiate into plasma cells, which are antibody-producing factories. Plasma cells are specialized for antibody secretion and can produce a vast quantity of antibodies specific to the antigen.

3.29. ANTIBODY PRODUCTION AND SECRETION

Plasma cells synthesize and secrete antibodies, which are structurally designed to recognize and bind to the specific antigen that triggered their production. Antibodies are Y-shaped proteins consisting of four polypeptide chains: two heavy chains and two light chains. The variable regions of the heavy and light chains form the antigen-binding site, which is highly specific to the antigen. Antibodies can neutralize pathogens by various mechanisms, including blocking their entry into host cells, promoting their destruction by immune cells, or triggering complement activation. The structure of an antibody is given in Figure 3.24.

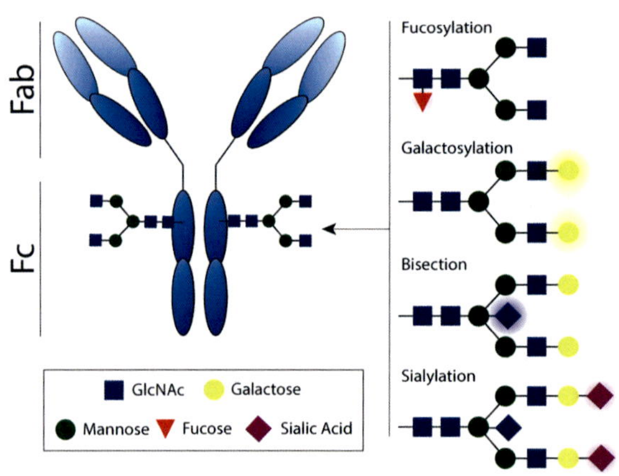

Figure 3.24. *The structure of an antibody.*

Key: Fab, antigen-binding fragment; Fc, crystallizable fragment; GlcNAc, N-acetylglycosamine.

Source: van Erp EA, Luytjes W, Ferwerda G and van Kasteren PB (2019) Fc-mediated antibody effector functions during respiratory syncytial virus infection and disease. Front. Immunol., 10, 548. doi: 10.3389/fimmu.2019.00548:

- **Memory B Cells:**After the immune response, a subset of activated B cells becomes memory B cells. Memory B cells "remember" the specific antigen, allowing for a faster and more robust response upon re-exposure to the same pathogen in the future.
- **Antibody-Mediated Immunity:**Antibodies provide immunity by recognizing and neutralizing pathogens. They can also participate in opsonization (marking pathogens for destruction by immune cells) and complement activation. Antibody-mediated immunity is highly specific and can confer long-lasting protection against previously encountered pathogens (Netea et al., 2016).
- **T Cell Activation:**T-cell activation is a critical step in the adaptive immune response, which helps the immune system recognize and combat specific pathogens or abnormal cells, such as infected cells or cancer cells. T cells, a type of white blood cell, play a central role in orchestrating and regulating the immune response. Here's a detailed explanation of T-cell activation:
- **Antigen Presentation:**T-cell activation begins with the presentation of antigens by antigen-presenting cells (APCs). APCs include dendritic cells, macrophages, and B cells. These APCs capture and process antigens from pathogens (e.g., proteins from viruses or bacteria) or abnormal cells (e.g., cancer cells). Antigens are broken down into smaller fragments and displayed on the surface of APCs using major histocompatibility complex (MHC) molecules.
- **T Cell Receptor (TCR) Recognition:** T cells express T cell receptors (TCRs) on their surface. TCRs are highly specialized proteins that can recognize specific antigen fragments presented on MHC molecules. When a TCR binds to its cognate antigen-MHC complex, this interaction is termed "antigen recognition." TCR binding is highly specific, and each T cell has a unique TCR.
- **Co-stimulation:** While TCR recognition is essential, it is not sufficient for full T-cell activation. Co-stimulatory signals are required to prevent inappropriate immune responses. One key co-stimulatory signal comes from the interaction between a protein on the T cell surface called CD28 and a molecule called CD80 or CD86 on the APC's surface. This co-stimulatory signal ensures that T cell activation only occurs when there is a genuine threat.

- **T Cell Activation:** Once the TCR recognizes the antigen-MHC complex, and co-stimulatory signals are received, the T cell becomes activated. T cell activation involves a series of intracellular signaling events, leading to changes in gene expression and functional responses.

A glimpse of this complex process has been shown in Figure 3.25.

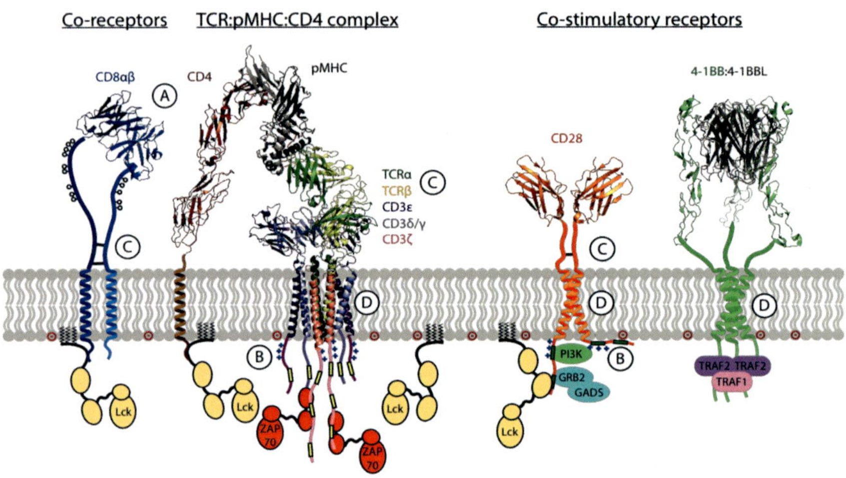

Figure 3.25.*A glimpse of complex process.*

Source: Chandler NJ, Call MJ, & Call ME, (2020). T cell activation machinery: Form and function in natural and engineered immune receptors. International Journal of Molecular Sciences, 21(19), 7424. https://doi.org/10.3390/ijms21197424: © 2020 by the authors. Licensee MDPI, Basel, Switzerland. This article is an open access article distributed under the terms and conditions of the Creative Commons Attribution (CC BY) license (http://creativecommons.org/licenses/by/4.0/).

- **Clonal Expansion:** Activated T cells undergo clonal expansion, which means they start to divide rapidly, producing a large population of identical T cells. This expansion ensures that there are enough T cells with the same antigen specificity to mount an effective immune response against the pathogen.
- **Differentiation:** T cells can differentiate into various functional subsets based on their specific role in the immune response. For example, some T cells become cytotoxic T cells (CD8+ T cells) that can directly kill infected or abnormal cells. Others become helper T cells (CD4+ T cells) that assist other immune cells by releasing cytokines, coordinating immune responses, and activating B cells.

- **Memory T Cells:** After the immune response subsides, a portion of the activated T cells differentiates into memory T cells. Memory T cells "remember" the antigen they encountered, providing long-term immunity. In future encounters with the same antigen, memory T cells can mount a faster and more robust response. T cell activation is a highly regulated process to ensure effective immune responses while avoiding unnecessary damage to healthy tissues. It plays a crucial role in adaptive immunity, working in concert with other immune cells like B cells and innate immune cells to protect the body from infections and abnormal cells (Nauseef&Borregaard, 2014).

3.30. MEMORY RESPONSES

A memory response, in the context of the immune system, refers to the rapid and robust immune reaction that occurs upon re-exposure to a pathogen or antigen (a molecule that the immune system recognizes as foreign). Memory responses are a fundamental aspect of adaptive immunity and are characterized by the immune system's ability to "remember" specific pathogens or antigens it has encountered in the past. These responses are faster, more efficient, and often more effective than the initial immune response to the same antigen. Here's a detailed explanation of memory responses:

- **Primary Immune Response:** When the immune system encounters a new pathogen or antigen for the first time, it initiates a primary immune response. The primary response involves the activation and proliferation of immune cells, such as B cells and T cells, that are specific to the encountered antigen. During the primary response, the immune system may take several days to generate an effective immune reaction, allowing the pathogen to cause disease symptoms.
- **Formation of Memory Cells:** As part of the primary immune response, some of the activated B cells and T cells differentiate into memory cells. Memory B cells and memory T cells are long-lived and remain in the body for an extended period, sometimes for a person's entire life. These memory cells "remember" the specific antigen or pathogen they encountered during the primary response.
- **Secondary Immune Response:** If the same pathogen or antigen re-enters the body at a later time, memory cells recognize it quickly and mount a secondary immune response. The secondary response is much faster and more efficient than the primary response because memory cells are already primed to respond. Memory B cells

quickly differentiate into plasma cells, which produce large amounts of antibodies specific to the antigen. Memory T cells also become activated and coordinate various immune responses.

3.31. CHARACTERISTICS OF MEMORY RESPONSES

- **Rapid:** Memory responses can occur within hours of re-exposure to the antigen, leading to a more immediate immune defense.
- **Stronger:** Memory responses typically generate a higher quantity of antibodies and activate more immune cells, ensuring a more effective defense against the pathogen.
- **Long-Lasting:** Memory cells persist in the body for an extended period, sometimes providing lifelong immunity to certain diseases.
- **Specific:** Memory responses are highly specific to the antigen or pathogen that triggered the memory cell formation. They do not respond to unrelated antigens.

1. **Types of Immunity:** Memory responses contribute to two main types of immunity:

i. **Active Immunity:** When memory cells are generated as a result of natural infection or vaccination, providing long-term protection.

ii. **Passive Immunity:** When pre-formed antibodies or immune cells from another individual (e.g., mother to fetus through breast milk) provide temporary protection without memory cell involvement.

3.32. INFLAMMATION

Inflammation is a complex and vital biological response of the body to various harmful stimuli, such as infections, tissue damage, or irritants. It plays a critical role in the body's defense mechanisms, helping to eliminate the cause of injury or infection and initiate the healing process. Inflammation involves a series of cellular and molecular events and is characterized by specific signs and symptoms. Here is a detailed explanation of inflammation:

1. Causes of Inflammation

i. **Infections:** Bacterial, viral, fungal, or parasitic infections trigger an immune response and inflammation.

ii. **Tissue Injury:** Physical trauma, burns, cuts, and chemical exposure can damage tissues and initiate inflammation.

iii. **Autoimmune Reactions:** In autoimmune diseases, the immune system mistakenly targets and attacks healthy tissues, leading to chronic inflammation.

iv. **Chronic Diseases**: Conditions like obesity, diabetes, and atherosclerosis can cause low-level, chronic inflammation (Netea et al., 2016).

2. Key Players in Inflammation

i. **Immune Cells:** White blood cells, particularly neutrophils and macrophages, are central to the inflammatory response. They detect and eliminate pathogens.

ii. **Cytokines:** Small signaling proteins that regulate immune responses and inflammation. Pro-inflammatory cytokines promote inflammation, while anti-inflammatory cytokines help resolve it.

iii. **Chemokines:** Attract immune cells to the site of inflammation.

iv. **Inflammatory Mediators:** Substances like histamines and prostaglandins are released and contribute to inflammation. They cause blood vessels to dilate and become more permeable (Figure 3.26).

The mechanisms of inflammation are summarized in Figure 3.26.

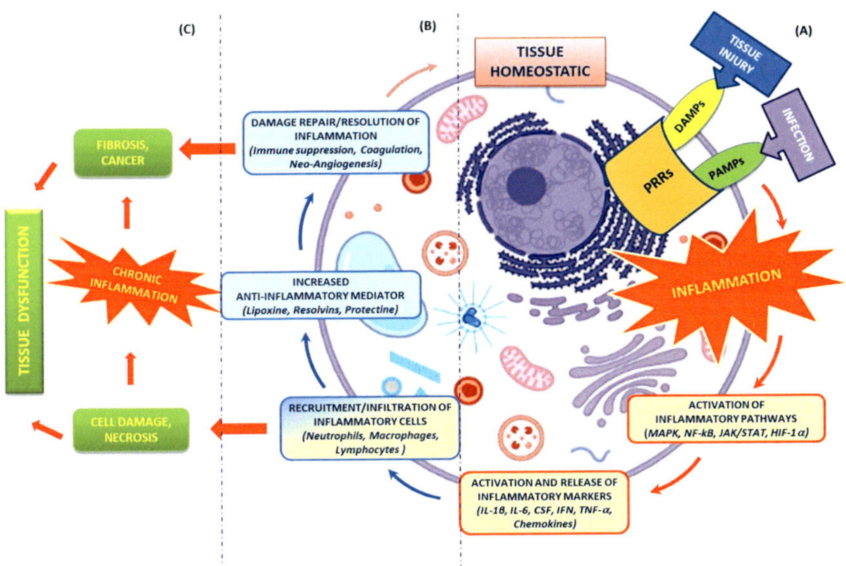

Figure 3.26.*The mechanisms of inflammation.*

Source: Nesci S, Spagnoletta A, &Oppedisano F, (2023). Inflammation, mitochondria and natural compounds together in the circle of trust. International Journal of Molecular Sciences, 24(7), 6106. https://doi.org/10.3390/ijms24076106:

3. Phases of Inflammation

i. **Vasodilation:** Blood vessels near the injured or infected site expand, allowing increased blood flow. This causes redness and warmth at the affected area (Woodfin et al., 2011).

ii. **Increased Vascular Permeability:** Blood vessels become leaky, leading to the accumulation of fluid, proteins, and immune cells in the tissues. Swelling or edema results.

iii. **Cellular Infiltration:** Immune cells, especially neutrophils, migrate to the site of inflammation to combat pathogens or remove damaged tissue.

iv. **Resolution and Repair:** Anti-inflammatory signals and mechanisms are activated to resolve inflammation and promote tissue healing.

4. Signs and Symptoms of Inflammation

i. **Redness (Rubor):** Due to increased blood flow and vasodilation.

ii. **Heat (Calor):** Also due to increased blood flow.

iii. **Swelling (Tumor):** Caused by fluid and immune cell accumulation.

iv. **Pain (Dolor):** Resulting from pressure on nerve endings and the release of inflammatory mediators.

v. **Loss of Function:** Inflammation can limit the function of the affected area, as seen in joint inflammation.

5. Acute vs. Chronic Inflammation

i. **Acute Inflammation:** Short-term response (hours to days) to injury or infection. It aims to eliminate the threat and promote healing.

ii. **Chronic Inflammation:** Persistent inflammation lasting weeks, months, or even years. It is often associated with underlying health conditions and can contribute to tissue damage and disease.

3.33. INFLAMMATORY DISEASES

- **Rheumatoid Arthritis:** An autoimmune disease characterized by chronic joint inflammation.
- **Inflammatory Bowel Disease (IBD):** Includes Crohn's disease and ulcerative colitis, causing chronic inflammation of the digestive tract.

- **Atherosclerosis:** Chronic inflammation plays a role in the development of arterial plaques.
- **Allergies:** Allergic reactions involve an inflammatory response.

Several chronic conditions and their association with inflammation are shown in Figure 3.27.

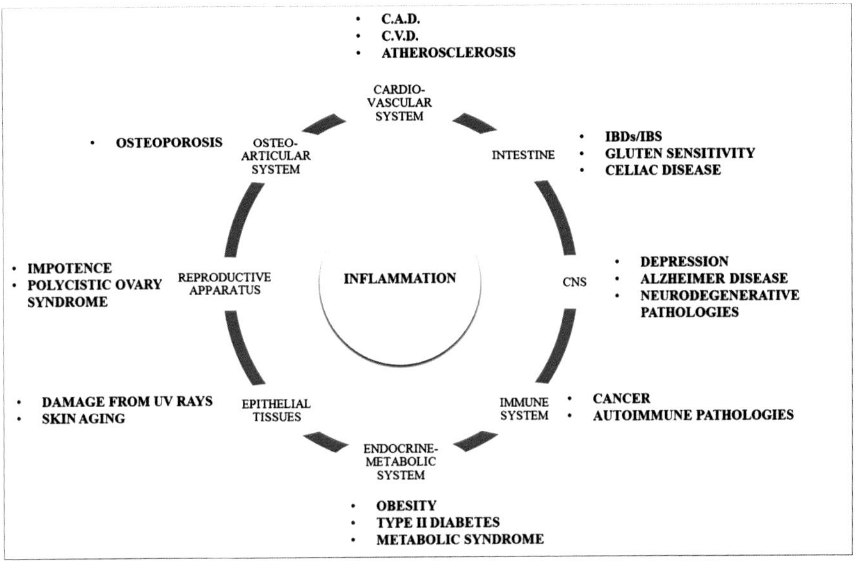

Figure 3.27.*Several chronic conditions and their association with inflammation.*

Source: Fioranelli M, Roccia MG, Flavin D, & Cota L, (2021). Regulation of inflammatory reaction in health and disease. International Journal of Molecular Sciences, 22(10), 5277. https://doi.org/10.3390/ijms22105277: © 2021 by the authors. Licensee MDPI, Basel, Switzerland. This article is an open access article distributed under the terms and conditions of the Creative Commons Attribution (CC BY) license (https://creativecommons.org/licenses/by/4.0/).

3.34. COMPLEMENT SYSTEM

The complement system is a complex and crucial part of the immune system that helps the body defend against infections, clear damaged cells, and regulate immune responses. It consists of a group of over 30 proteins that work together to enhance the immune response. The complement system plays a vital role in both the innate and adaptive immune responses. Here is a detailed explanation of the complement system (Nauseef&Borregaard, 2014).

3.35. Complement Proteins

The complement system comprises various proteins, including:

- **Complement Proteins (C1-C9):** These are numbered from C1 to C9 and include proteins like C3, C4, and C5.
- **Regulatory Proteins:** These proteins control and regulate the complement system to prevent excessive activation.
- **Receptor Proteins:** These are present on the surfaces of immune cells and help mediate the effects of complement activation.

3.36. ACTIVATION PATHWAYS

- **Classical Pathway:** This pathway is primarily triggered by antibodies bound to antigens on the surface of pathogens or other cells. It involves a sequence of complement proteins, including C1, C4, and C2.

The activation of the complement system has been depicted in Figure 3.28.

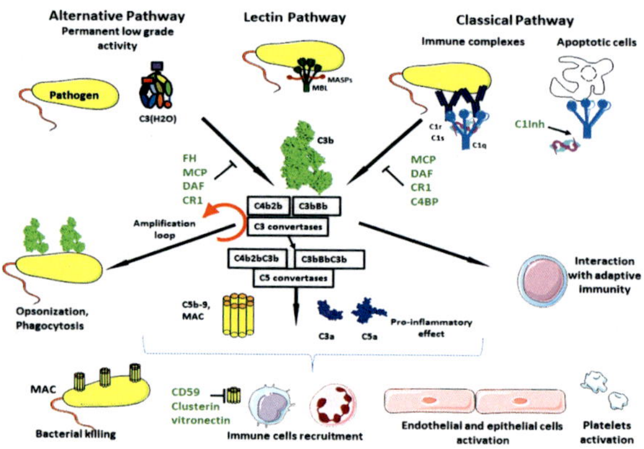

Figure 3.28. *The activation of the complement system.*

Source: Merle NS, Noe R, Halbwachs-Mecarelli L, Fremeaux-Bacchi V &Roumenina LT, (2015) Complement system part II: role in immunity. Front. Immunol., 6, 257. doi: 10.3389/fimmu.2015.00257: Copyright: © 2015 Merle, Noe, Halbwachs-Mecarelli, Fremeaux-Bacchi and Roumenina. This is an open-access article distributed under the terms of the Creative Commons Attribution License (CC BY). The use, distribution or reproduction in other forums is permitted, provided the original author(s) or licensor are credited and that the original publication in this journal is cited, in accordance with accepted academic practice. No use, distribution or reproduction is permitted which does not comply with these terms.

- **Alternative Pathway:** This pathway is activated directly by the presence of pathogens or foreign substances. It involves factors like properdin and factor B.
- **Lectin Pathway:** This pathway is initiated when pattern-recognition molecules, such as mannose-binding lectin (MBL), bind to specific carbohydrates on pathogens. It activates complement proteins like C4 and C2.

3.37. COMPLEMENT ACTIVATION STEPS

- **Recognition:** The complement system is activated when certain molecules, such as antibodies, microbial cell surfaces, or lectins, are recognized as foreign or dangerous.
- **Initiation**: In the classical pathway, the C1 complex (C1q, C1r, and C1s) binds to antigen-antibody complexes. In the alternative pathway, factors like properdin bind to microbial surfaces.
- **Amplification:** Once initiated, the cascade of complement protein activations amplifies the response, leading to the activation of downstream proteins.
- **Cascade:** The activation of one complement protein leads to the activation of others through proteolytic cleavage, forming complexes that enhance immune responses.
- **Opsonization:** Complement proteins can coat pathogens, making them more recognizable by phagocytic cells like macrophages and neutrophils.
- **Inflammation:** Some complement proteins can trigger inflammation by attracting immune cells to the site of infection or injury.
- **Membrane Attack Complex (MAC):** The terminal steps of complement activation result in the formation of the MAC, which creates pores in the target cell's membrane, leading to cell lysis.

3.38. FUNCTIONS OF THE COMPLEMENT SYSTEM

- **Opsonization**: Enhances phagocytosis by labeling pathogens for ingestion by immune cells (Woodfin et al, 2011).
- **Inflammation:** Helps recruit immune cells to the site of infection or injury.
- **Cell Lysis:** The MAC formed by complement proteins can puncture and destroy the membranes of target cells, such as bacteria.

- **Clearance of Immune Complexes:** Helps clear immune complexes from circulation.
- **Enhancement of Adaptive Immunity:** Complement can enhance antibody production and the adaptive immune response.

3.39. REGULATION OF COMPLEMENT ACTIVATION

- **Regulatory Proteins:** Factors like C1-inhibitor, decay-accelerating factor (DAF), and membrane cofactor protein (MCP) prevent excessive complement activation on host cells.
- **Soluble Regulators:** Proteins like factor H and factor I control complement activation in body fluids.
- **C3 Convertase Regulation:** DAF and MCP inhibit C3 convertase formation.

3.39.1. Clinical Implications

- **Deficiencies:** Complement deficiencies can lead to an increased susceptibility to infections, autoimmune diseases, or immune complex diseases.
- **Therapeutic Use:** Some complement inhibitors are used in the treatment of diseases like paroxysmal nocturnal hemoglobinuria (PNH) and certain autoimmune disorders.

The complement system is a critical component of the immune response, working in concert with other immune mechanisms to protect the body from infections and maintain immune homeostasis. Dysregulation of the complement system can have serious consequences for health, making it an important area of study and clinical interest.

3.40. WHITE BLOOD CELLS IN HEALTH AND DISEASE: APPLICATION AND IMPLICATIONS

White blood cells are essential components of the immune system with applications spanning diagnosis, therapy, and research. Understanding their roles and behaviors in health and disease is critical for developing effective treatments, managing infections, and advancing our knowledge of immunology and human biology. Let us now understand the different interesting aspects of the application and implications of white blood cells.

1. **WBCs as Diagnostic Biomarkers:** White blood cells (WBCs) serve as important diagnostic markers in various medical conditions

and diseases. Changes in WBC counts and differentials (the percentages of different types of WBCs in the blood) can provide valuable information to healthcare professionals for diagnosing and monitoring a wide range of health issues. Here's how WBCs are used as diagnostic markers.

2. **Infection Detection**
   i. **Bacterial Infections:** An elevated total WBC count (leukocytosis), particularly with increased neutrophils (neutrophilia), often indicates a bacterial infection. Neutrophils are the first responders to bacterial invaders (Woodfin et al, 2011).
   ii. **Viral Infections:** Some viral infections, especially those involving lymphocytes (e.g., mononucleosis or Epstein-Barr virus), can lead to an increase in lymphocyte counts (lymphocytosis).
   iii. **Parasitic and Fungal Infections:** Certain parasitic and fungal infections may cause an increase in eosinophils or basophils, depending on the specific infection.
3. **Inflammatory Conditions:**
   i. **Chronic Inflammatory Diseases:** Conditions like rheumatoid arthritis, lupus, and inflammatory bowel disease can lead to persistent leukocytosis.
   ii. **Acute Inflammation:** Acute inflammatory conditions, such as appendicitis, can cause a rise in total WBC count due to the body's immune response.
4. **Blood Disorders:**
   i. **Leukemia:** Abnormal WBC counts, particularly very high total WBC counts and immature cells (blasts), can be indicative of leukemia or other bone marrow disorders.
   ii. **Lymphoma:** Specific patterns of lymphocyte counts can help distinguish between Hodgkin's and non-Hodgkin's lymphomas.
5. **Immunodeficiency and Autoimmune Diseases:**
   i. **HIV/AIDS:** Monitoring CD4+ T cell counts is crucial in managing HIV infection. A decrease in CD4+ T cell counts suggests disease progression.

ii. **Autoimmune Diseases:** Aberrant WBC counts and differentials are often seen in autoimmune diseases like systemic lupus erythematosus (SLE) and rheumatoid arthritis.

6. **Allergies and Asthma:**

Allergies*:* Increased eosinophil counts can be seen in allergic conditions such as asthma and allergic rhinitis.

Asthma*:* Elevated eosinophils and basophils may indicate an asthma exacerbation.

7. **Monitoring Treatment:**

i. **Chemotherapy:** WBC counts are monitored in cancer patients receiving chemotherapy to assess the impact on bone marrow function.

ii. **Immune Suppression:** Transplant recipients and patients on immunosuppressive drugs are monitored for low WBC counts, which can indicate an increased risk of infection.

8. **Hematological Disorders**

i. **Myeloproliferative Disorders:**Conditions like polycythemia vera and essential thrombocythemia can affect WBC counts.

ii. **Myelodysplastic Syndromes:** Abnormalities in WBC counts and differentials can be an early sign of myelodysplastic syndromes.

9. **Infection Severity**

i. **COVID-19:** Monitoring changes in WBC counts, particularly lymphocyte levels, can help assess the severity of COVID-19 infection (Zhang et al, 2015).

10. **Risk Assessment**

i. **Cardiovascular Risk:** High WBC counts are associated with an increased risk of cardiovascular events, including heart attacks and strokes.

11. **Bone Marrow Disorders**

i. **Aplastic Anemia:** A marked decrease in all blood cell types, including WBCs, is seen in aplastic anemia.

Thus, changes in WBC counts and differentials are valuable diagnostic markers that can provide insights into various health conditions. Healthcare providers use this information, along with other clinical and laboratory data, to make accurate diagnoses, monitor disease progression, and determine appropriate treatment strategies (Nauseef&Borregaard, 2014).

3.41. DISEASE-SPECIFIC SIGNATURES

Disease-specific signatures, often referred to as molecular or biomolecular signatures, are patterns or profiles of molecules, such as genes, proteins, or metabolites, that are associated with a particular disease or medical condition. These signatures are identified through various omics technologies (genomics, proteomics, metabolomics, etc.) and are valuable for diagnosing diseases, predicting disease outcomes, developing targeted therapies, and understanding disease mechanisms. Here, we'll explain disease-specific signatures in detail (Zhang et al., 2015).

1. Genomic Signatures

i. **Gene Expression Profiles:** In diseases like cancer, specific genes can be overexpressed or under-expressed, leading to characteristic gene expression patterns. Microarray or RNA sequencing technologies are used to identify these profiles.

ii. **Genetic Mutations:** Diseases like cystic fibrosis, sickle cell anemia, and certain cancers have well-defined genetic mutations. Detection of these mutations in an individual's DNA serves as a disease-specific signature.

iii. **Epigenetic Changes:** Aberrant DNA methylation or histone modification patterns are associated with diseases such as cancer. Epigenetic signatures reflect changes in gene regulation.

As an example, research findings by Li et al. (2023) have been shared below. The study entailed the analysis of immune cells by Gene Set Enrichment Analysis (GSEA) and Tumor Immune Estimation Resource (TIMER). Many pathways were identified to be expressed differently in the blood cells of patients with sepsis vs. controls (Figure 3.29).

2. Proteomic Signatures

i. **Protein Expression Profiles:** Proteomic studies identify disease-specific protein expression patterns. Mass spectrometry and antibody-based assays are used to analyze proteins.

ii. **Post-Translational Modifications:** Disease-specific protein modifications, such as phosphorylation or glycosylation, can be detected as signatures. These modifications may impact protein function.

iii. **Biomarkers:** Specific proteins can serve as biomarkers for diseases. For example, elevated levels of prostate-specific antigen (PSA) are a signature of prostate cancer (Woodfin et al., 2011).

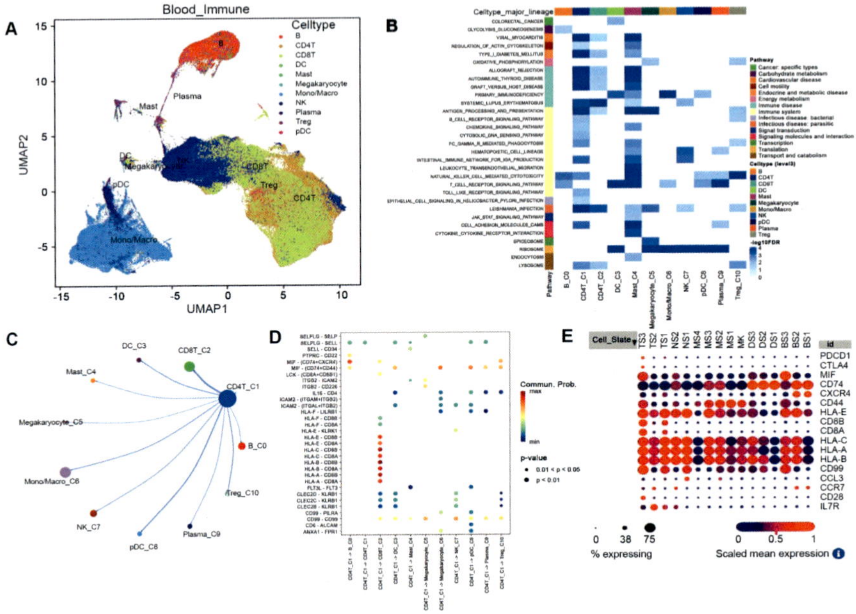

Figure 3.29. Pathways identified to be expressed differently in the blood cells of patients with sepsis vs. controls in the Li et al (2023) study

Key: (A) Uniform Manifold Approximation and Projection (UMAP)plot for the 11 immune cell clusters. (B) Heatmap showing down-regulated pathways. (C) Circular plot for cell communications between CD4⁺ T cell and other 10 immune cells. (D) All the significant ligand-receptor pairs involved in signaling. (E) T cell exhaustion; TS, T cell subclusters; NS, NK cell subclusters; MS, monocyte/macrophage subclusters; MK, megakaryocyte subclusters; DS, dendritic cell (DC) subclusters; BS, B cell subclusters.

Source: Li Q, Sun M, Zhou Q, Li Y, Xu J and Fan H (2023) Integrated analysis of multi-omics data reveals T cell exhaustion in sepsis. Front. Immunol., 14, 1110070. doi: 10.3389/fimmu.2023.1110070:

3. Metabolomic Signatures

i. **Metabolite Profiles:** Metabolomics focuses on small molecules (metabolites) involved in cellular processes. Altered metabolite profiles can indicate disease states, as seen in metabolic disorders.

ii. **Metabolic Pathways:** Dysregulation of metabolic pathways, such as glycolysis or the citric acid cycle, can produce disease-specific signatures. These pathways are assessed using metabolomics.

4. Microbiome Signatures

i. **Gut Microbiota:** The composition of gut microbiota is associated with various diseases, including inflammatory bowel disease (IBD) and obesity. Microbiome signatures reflect the types and abundance of microbial species (Zhang et al., 2015).

ii. **Oral Microbiome:** Dental and oral diseases have unique microbiome signatures characterized by shifts in microbial populations.

5. Imaging Signatures

i. **Medical Imaging:** Radiological and imaging techniques, like MRI, CT scans, and PET scans, reveal structural and functional changes in tissues and organs. Patterns observed in images can be diagnostic signatures.

6. Clinical Signatures

i. **Symptom Clusters:** Disease-specific clusters of symptoms, such as those seen in neurological disorders, can serve as clinical signatures.

ii. **Diagnostic Criteria:** Certain diseases are defined by specific criteria outlined in medical guidelines, and meeting these criteria is a diagnostic signature.

7. Immune Signatures

i. **Immune Cell Profiles:** Diseases often result in immune system responses. Immune cell populations, cytokine levels, and antibody titers can serve as immune signatures.

ii. **Autoantibodies:** Autoimmune diseases are characterized by the presence of autoantibodies against self-antigens, forming autoimmune signatures.

3.42. IMMUNOTHERAPY: TARGETED TREATMENTS

Immunotherapy is a cutting-edge approach to treating various diseases, particularly cancer and autoimmune disorders, by harnessing the power of the immune system, including white blood cells, to target and eliminate diseased cells. Here's a detailed explanation of immunotherapy and how it relates to white blood cells:

3.42.1. Types of Immunotherapies

Several types of immunotherapies are shown in Figure 3.30.

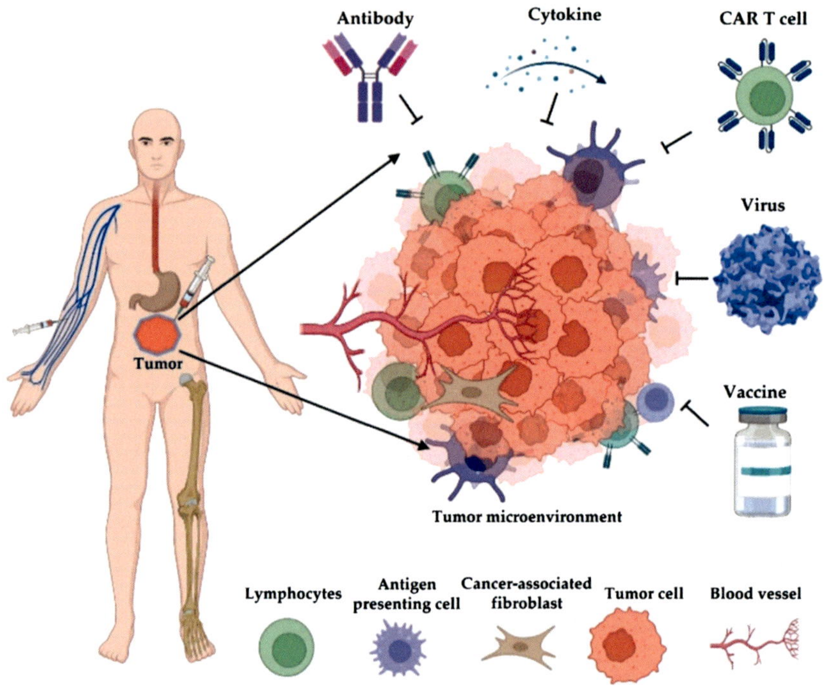

Figure 3.30. *Types of immunotherapies.*

Source: Yang M, Olaoba OT, Zhang C, Kimchi ET, Staveley-O'Carroll KF, Li G, (2022). Cancer immunotherapy and delivery system: An update. Pharmaceutics, 14(8), 1630. https://doi.org/10.3390/pharmaceutics14081630: © 2022 by the authors. Licensee MDPI, Basel, Switzerland. This article is an open access article distributed under the terms and conditions of the Creative Commons Attribution (CC BY) license (https://creativecommons.org/licenses/by/4.0/).

- **Checkpoint Inhibitors:** Checkpoint inhibitors are drugs that block certain proteins, like PD-1 and CTLA-4, on the surface of T cells. By doing so, they "release the brakes" on the immune system, allowing it to attack cancer cells more effectively.

Table 3.8. Select CAR therapies undergoing clinical trials (as of 2022)

Disease Condition	Drug Combinations	Phase	Participants	Status	NCT Number
Breast cancer	4SCAR T cells (CAR-T cells targeting Her2, GD2, and CD44v6)	I/II	100	Recruiting	NCT04430595
Acute myeloid leukemia	CAR-T CD19 CD7 CAR-T cells	II/III I/II	10 108	Recruiting Recruiting	NCT04257175 NCT04599556
Multiple myeloma	CAR-T cell targeting B-cell maturation antigen (BCMA), Bortezomib, Dexamethasone, Lenalidomide, Cyclophosphamide, Fludarabine. JNJ-68284528 (cilta-cel), Pomalidomide, Bortezomib, Dexamethasone, Daratumumab JNJ-68284528 (ciltacabtageneautoleucel [cilta-cel], Bortezomib, Lenalidomide, Dexamethasone, Cyclophosphamide, Fludarabine, Daratumumab	III III III	650 419 750	Recruiting Active not yet recruiting Not yet recruiting	NCT04923893 NCT04181827 NCT05257083

B cell lymphoma	CAR-T-CD19, BTK inhibitor, Fludarabine, Cyclophosphamide	III	24	Recruiting	NCT05020392
B Cell malignancies	CD19/CD22-CAR-T cells, fludarabine, cyclophosphamide	I/II	146	Not yet Recruiting	NCT05442515
Solid tumor	CLDN6 CAR-T, CLDN6 RNA-LPX	I/II	96	Recruiting	NCT04503278
Pancreatic cancer	CD276 CAR-T cells	I/II	10	Recruiting	NCT05143151
Gastric cancer, Pancreatic cancer	CT041 (CAR-T cells targeting claudin18.2)	I/II	110	Recruiting	NCT04404595
Prostate cancer	4SCAR-PSMA T cells [CAR-T cells targeting Prostate-specific membrane antigen (PSMA)]	I/II	100	Recruiting	NCT04429451
CD44v6 positive cancers (squamous cell carcinomas, adenocarcinomas, melanoma, lymphoma)	4SCAR-CD44v6 [CAR-T cells targeting CD44v6]	I/II	100	Recruiting	NCT04427449

Source: Mishra AK, Ali A, Dutta S, Banday S, &Malonia SK, (2022). Emerging trends in immunotherapy for cancer. Diseases, 10(3), 60. https://doi.org/10.3390/diseases10030060:

- **CAR T-Cell Therapy:** Chimeric Antigen Receptor (CAR) T-cell therapy involves genetically engineering a patient's T cells to express a receptor that recognizes a specific antigen in cancer cells. These modified T cells are then infused back into the patient to target and kill cancer cells (Zhang et al., 2015). Mishra et al. (2022) summarized select CAR therapies undergoing clinical trials (ClinicalTrials.gov, www.clinicaltrials.gov, as accessed in June 2022).
- **Monoclonal Antibodies:** Monoclonal antibodies are lab-made proteins that can mimic the immune system's ability to fight off harmful cells. They can be designed to target specific proteins or antigens on the surface of cancer cells, thereby marking them for destruction. In a comprehensive review by an update from the World Association of Infectious Diseases and Immunological Disorders and the Italian Society of Antinfective Therapy, Esposito et al. (2023) summarized monoclonal antibodies against viral diseases. It has been presented in Table 3.9.

Table 3.9. Monoclonal antibodies against viral diseases

	Action	Effect/ Efficacy	Indications	Dose	Half-Life	Cost	Potential Use
RSV prevention							
Palivizumab (23–36)	Humanized mAB Target RSV F protein, site II	45-82% against RSV related hospitalizations in high risk infants	Infants <29 weeks of gestational age; chronic lung disease of prematurity; hemodynamically significant congenital heart defect; selected cases of pulmonary abnormalities, neuromuscular disorders and severe immunodeficiency	15 mg/kg once a month (max 5) during RSV season	Short (17-26 days)	Very high	Reduced (high risk groups only)
Nirsevimab (40–51)	Humanized mAB Target RSV prefusion F protein, site ϕ	Preterms: 70,1% reduction on RSV-associated MALRTIs, 2,6% vs 9,5% on RSV disease compared to placebo and 78,4% reduction on hospitalization rate due to RSV-associated MALRTIs. Preterm and Term Infants: 74,5% reduction on RSV-associated MALRTIs and 62,1% reduction on hospitalization	Universal indication (potential) on infants	Single dose: 50 mg IM in <5 kg and 100 mg IM in > 5kg	Extended half-life due to modified Fc region (≈150 days)	Potential favorable cost-effectiveness ratio	Universal use of RSV prophylaxis in children
Clesrovimab (52)	Humanized mAB Target RSV prefusion F protein, site IV	Studies ongoing					
SARS-CoV-2 prophylaxis or therapy							
Bebtelovimab (66–73)	Human IgG1 mAB Target SARS-CoV-2 S-protein. Active against Omicron BA.5 and Omicron BA.4.6/BF.7	Reduction on viral load on day 5 after treatment, compared to placebo. Time to sustained symptom resolution was reduced by a median of 2 days	Emergency use in adults and children >12 years of age and weighting >40 kg. Not authorized in hospitalized patients	175 mg IV	N/A	High	Patients with mild to moderate disease at high risk to progression.
Tixagevimab plus cilgavimab (78, 79)	Active against Omicron BA.5.	*In vivo* efficacy clearly evidenced only in patients infected by non-Omicron variants. i	Emergency use in adults and children >12 years of age and weighting >40 kg for pre-exposure prophylaxis in immunocompromised individuals or those who cannot be vaccinated or mount post-vaccination immune response	300 mg + 300 mg IM	N/A	Hugh	Post-exposition prophylaxis in selected patients with infection due to sensitive

	Action	Effect/ Efficacy	Indications	Dose	Half-Life	Cost	Potential Use
							SARS-CoV-2 strains
HIV therapy							
Ibalizumab (81–83)	Humanized mAB Target CD4 T cell extracellular domain 1 and 2	*In vitro* neutralizing activity against approximately 90% of a diverse panel of HIV strains. Very fast reduction of HIV-RNA levels. After 24 weeks of treatment 43% of participants achieve HIV RNA suppression	In adults as part of a combination antiretroviral regimen in heavily treatment-experienced patients with multidrug resistant (MDR) HIV-1 infection who did not respond to the current antiretroviral regimen	First dose: 2000 mg IV; maintenance doses 800 mg every 2 weeks	Extended half-life with high dose due to saturable elimination	Very high	Only in patients unresponsive to antivirals
Rabies prevention							
Rabishield	Target a conformational epitope of the rabies G protein	It is as effective as serum derived hyperimmune IgG. It can fail against virus variants tat circulate in Africa and North America	It must be given together with vaccine within 7 days after exposure to rabies virus	3.33 IU/kg body weight		Less expensive than hyperimmune rabies IgG	Less recommended than preparations with two mABs
Miromavimab plus Docaravimab	Target the antigenic sites I and III of the rabies G protein	Combination is as effective as serum derived hyperimmune IgG	It must be given together with vaccine within 7 days after exposure to rabies virus	40 IU/kg body weight		Less expensive than hyperimmune rabies IgG	Extensive use

N/A, not applicable.

Source: Esposito S, Amirthalingam G, Bassetti M, Blasi F, De Rosa FG, Halasa NB, Hung I, Osterhaus A, Tan T, Torres JP, Vena A &Principi N, (2023) Monoclonal antibodies for prophylaxis and therapy of respiratory syncytial virus, SARS-CoV-2, human immunodeficiency virus, rabies and bacterial infections: an update from the World Association of infectious diseases and immunological disorders and the Italian Society of antinfective therapy. Front. Immunol., 14, 1162342. doi: 10.3389/fimmu.2023.1162342:

- **Cytokine Therapy:** Cytokines are signaling molecules that regulate immune responses. In some cases, patients are given cytokines, such as interleukins or interferons, to boost the immune system's activity against cancer.
- **Vaccines:** Cancer vaccines are designed to stimulate the immune system to recognize and target cancer cells. These vaccines can be composed of cancer-specific antigens or tumor cells themselves.
- **Immune Checkpoint Blockade:** Immune checkpoint molecules like PD-1 and PD-L1 can inhibit immune responses against cancer.

Drugs that block these checkpoints are used to enhance the immune response.

A few FDA-approved immune checkpoint inhibitors (CPIs) compiled by Basudan (2022) are given in Table 3.10.

Table 3.10. A few FDA-approved immune checkpoint inhibitors (CPIs)

CPI Target	Drug	Class	1st Approval Year	Indication(s)
CTLA-4	Ipilimumab	IgG1 mAb	2010	Melanoma.
PD-1	Pembrolizumab	IgG4 mAb	2016	Melanoma, NSCLC, HNSCC, MSI-H solid tumors, GC, cervical cancer, urothelial carcinoma, TNBC, TMB-H/dMMR endometrial carcinoma.
	Nivolumab	IgG4 mAb	2014	Melanoma, NSCLC, RCC, Hodgkin's lymphoma, HNSCC, HCC, ESCC, pleural mesothelioma, MSI-H/dMMR CRC.
	Cemiplimab	IgG4 mAb	2018	Cutaneous squamous cell carcinoma.
PD-L1	Atezolizumab	IgG1 mAb	2016	Urothelial carcinoma, NSCLC, SCLC, melanoma, HCC.
	Durvalumab	IgG1 mAb	2017	Urothelial bladder cancer, NSCLC, SCLC, BTC.
	Avelumab	IgG1 mAb	2017	Merkel cell carcinoma, urothelial carcinoma, RCC.

Key: CTLA-4 = Cytotoxic T-lymphocyte-associated protein 4; PD-1/PD-L1 = Programmed cell death protein-1 and its ligand-1, respectively; IgG = Immunoglobulin

G; NSCLC = Non-small-cell lung cancer; HNSCC = Squamous cell carcinoma of the head and neck; GC = Gastric cancer; TNBC = Triple-negative breast cancer; RCC = Renal cell carcinoma; HCC = Hepatocellular carcinoma; ESCC = Esophageal squamous cell carcinoma; CRC = Colorectal cancer, SCLC = Small-cell lung cancer; BTC = Biliary tract cancer; TMB-H = Tumors with high mutational burden; MSI-H = high microsatellite instability; dMMR = Mismatch repair deficient.

Source: Basudan AM, (2023). The role of immune checkpoint inhibitors in cancer therapy. Clinics andPractice, 13(1), 22–40. https://doi.org/10.3390/clinpract13010003:

- **Tumor-Infiltrating Lymphocytes (TILs):** TILs are white blood cells, mainly T cells, that have infiltrated the tumor. They can be harvested from the patient, expanded in the lab, and then infused back into the patient to target cancer cells.
- **Oncolytic Viruses:** These are viruses that are engineered to infect and kill cancer cells while also stimulating an immune response. They can activate white blood cells to attack the tumor (Zhang et al., 2015).

3.43. WHITE BLOOD CELLS IN IMMUNOTHERAPY

- **T Cells:** CAR T-cell therapy and TIL therapy involve harvesting the patient's T cells, modifying them to recognize cancer cells, and then infusing them back into the patient. These modified T cells can specifically target and kill cancer cells.
- **Antibodies:** Monoclonal antibodies can be designed to target specific antigens on cancer cells. Once bound to these cells, they can recruit immune cells, including white blood cells, to attack the cancer.
- **Immune Activation:** Some immunotherapies work by activating the patient's own immune cells, including white blood cells, to increase their ability to recognize and destroy cancer cells.
- **Immune Memory:** Immunotherapy can also help establish immune memory, wherein the immune system "remembers" the cancer cells and is better equipped to respond if the cancer returns (Woodfin et al, 2011).

3.44. CHALLENGES AND SIDE EFFECTS

While immunotherapy has shown great promise, it can also lead to immune-related side effects, where the immune system may attack normal healthy tissues. Managing these side effects is an important aspect of treatment (Nauseef&Borregaard, 2014).

Not all patients respond to immunotherapy, and research is ongoing to understand why some individuals benefit while others do not.

Immunotherapy is a rapidly evolving field with the potential to revolutionize the treatment of cancer and autoimmune diseases. By leveraging white blood cells and the immune system's natural capabilities, it offers targeted and precise treatments that can lead to improved outcomes and fewer side effects compared to traditional therapies. However, it also presents challenges that require ongoing research and development.

3.45. TRANSPLANTATION AND IMMUNOSUPPRESSION

White blood cells (WBCs) play crucial roles in transplantation and immunosuppression, primarily in the context of organ transplantation. Here's an explanation of how WBCs are applied in these areas:

1. ***Transplantation and White Blood Cells***

i. **Immunological Challenge:** Organ transplantation involves replacing a damaged or failing organ with a healthy one, which can trigger a significant immune response in the recipient's body. The immune system recognizes the transplanted organ as foreign (non-self) and mounts an attack to reject it.

i. **Role of White Blood Cells:** WBCs, including T cells, B cells, and various immune cell subsets, are central to the immune response against the transplanted organ. They are responsible for detecting and targeting the foreign antigens present in the donor organ.

2. Immunoregulation and Immunosuppression

i. **Immunosuppression:** To prevent organ rejection, transplant recipients often receive immunosuppressive therapy. Immunosuppressive drugs are designed to dampen the recipient's immune response, making it less likely to attack the transplanted organ.

3. WBCs in Immunosuppression

i. **T Cell Suppression:** Many immunosuppressive drugs target T cells, a type of WBC, as they are central to the rejection process. Drugs like calcineurin inhibitors (e.g., cyclosporine, tacrolimus) and mTOR inhibitors (e.g., sirolimus, everolimus) inhibit the activation and proliferation of T cells, reducing their ability to attack the transplant.

ii. B Cell Suppression: B cells, another type of WBC, can produce antibodies that target the transplant. Medications like rituximab, a monoclonal antibody, can selectively deplete B cells and reduce antibody-mediated rejection.

iii. **Anti-Inflammatory Effects:** Some immunosuppressive drugs have anti-inflammatory properties that affect various types of WBCs, reducing overall immune activity (Zhang et al., 2015).

As an example, a summary of factors that function in immunosuppression in colorectal cancer as reviewed by Zhang et al. (2020) is presented in Table 3.11.

Table 3.11. A Summary of Factors that Function in Immunosuppression in Colorectal Cancer

Classes	Immunosuppressors	Major Immune Cells Affected
Cell surface protein	PD-1	T cell, NK cell, TAM, DC
	TIGIT	T cell, NK cell
	TGFβ RII	Treg
	GARP	CD4+ and CD8+ T cells
	CD39	NK cell
	CD73	Effector T cell
	A2A adenosine receptors	T cell
	CCR8	Treg
	Disrupted Immunosurveillance Contributors	Major Immune Cells Affected
	NKG2D receptor	NK cell
	TLRs	Innate immune cell
	CD80	T cell

Cytokines/ Chemokines	Immunosuppressors	Major Immune Cells Affected
	TGFβ	Treg
	VEGF	T cell, MDSC, DC
	IL-6	CD4+ and CD8+ T cells
	CXCL3 and CXCL4	Treg, CD8+ T cell
	HMGB1	PMDDC
	Disrupted Immune Activators	Major Immune Cells Affected
	IL-12	NK cell, T cell
	IL-15	NK cell, CD8+ T cell
	IL-24	CD4+ and CD8+ T cells
	CXCL9, CXCL10, and CXCL11	CD4+ and CD8+ T cells
Phosphatase	Immunosuppressor	Major Immune Cells Affected
	SHP2	CD4+ and CD8+ T cells
Transcriptional factors	Immunosuppressors	Major Immune Cells Affected
	PRC2	Th1 and CD8+ T cells
	FOXM1	BMDC
	NF-κB	DC
	STAT1	CD4+ T cell
	STAT3	CD4+ T cell, DC
	TRF2	NK cell
Metabolic enzymes	Immunosuppressors	Major Immune Cells Affected
	ODC	M1 macrophage
	IDO1	CD4+ and CD8+ T cells, NK cell, MDSC
Other factors	Immunosuppressive factors	Major Immune Cells Affected
	Extracellular adenosine level	T cell
	Hypoxia	Treg, CD4+ effector T cell, MDSC, TAM, DC

Source: Zhang Y, Rajput A, Jin N, & Wang J, (2020). Mechanisms of immunosuppression in colorectal cancer. Cancers, 12(12):3850. https://doi.org/10.3390/cancers12123850: © 2020 by the authors. Licensee MDPI, Basel, Switzerland.

Key: dendritic cell, DC; tyrosine-based inhibitory motif domain, TIGIT; high mobility group box-1, HMGB1; peritoneal macrophage-derived dendritic cells, PMD-DC; ornithine decarboxylase, ODC

3.46. WHITE BLOOD CELL MONITORING

- **Routine Monitoring:** Transplant recipients undergo routine blood tests to monitor WBC counts and their subsets. Elevated WBC counts or changes in specific subsets can indicate potential issues with immunosuppression or organ rejection.
- **Detecting Rejection:** Changes in WBC counts, especially T cell counts, can signal acute rejection. Regular monitoring allows for early detection and adjustment of immunosuppressive therapy.
- **Infections:** Immunosuppression increases the risk of infections. Monitoring WBC counts helps identify infections promptly so that they can be treated without compromising the transplanted organ.
- **Leukocyte Depletion:**In some cases, especially in stem cell or bone marrow transplantation, leukocyte (WBC) depletion may be required. This process involves removing WBCs from the donor's graft before transplantation to reduce the risk of graft-versus-host disease (GVHD), a condition where donor WBCs attack the recipient's tissues (Nauseef&Borregaard, 2014).

3.47. CHALLENGES AND BALANCING ACT

- **Balancing Act:** Achieving the right balance of immunosuppression is crucial. Too much immunosuppression can lead to infections and other complications, while too little can result in organ rejection.
- **Targeted Therapies:** Ongoing research aims to develop more targeted immunosuppressive therapies that affect specific subsets of white blood cells or immune pathways, minimizing side effects.

3.48. FUTURE DIRECTIONS

- **Personalized Medicine:**The field of transplantation is moving towards personalized immunosuppression, where treatment is tailored to the individual patient based on their immune profile and risk factors.

In summary, white blood cells play a pivotal role in organ transplantation and immunosuppression. The careful management of these cells, through monitoring and targeted immunosuppressive therapies, is essential to ensure the success of transplantation while minimizing the risk of rejection and other complications. Ongoing research continues to refine these approaches to improve outcomes for transplant recipients.

The use of the latest approaches like OMICs is vital in the field of precision medicine, as shown in Figure 3.31.

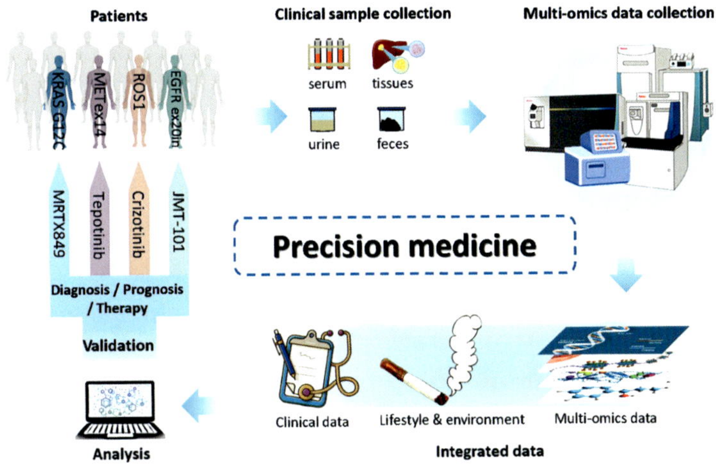

Figure 3.31. The roles of OMICs in precision medicine

Source: Su M, Zhang Z, Zhou L, Han C, Huang C, & Nice EC, (2021). Proteomics, personalized medicine and cancer. Cancers, 13(11), 2512. https://doi.org/10.3390/cancers13112512: © 2021 by the authors. Licensee MDPI, Basel, Switzerland. This article is an open access article distributed under the terms and conditions of the Creative Commons Attribution (CC BY) license (https://creativecommons.org/licenses/by/4.0/).

3.49. AUTOIMMUNE DISEASE MANAGEMENT

White blood cells (WBCs) play a significant role in the management of autoimmune diseases, which are conditions where the immune system mistakenly attacks the body's own tissues. Here's how WBCs are involved in autoimmune disease management:

- Immune Dysregulation Detection
    - **Autoimmune Diseases:** In autoimmune diseases like rheumatoid arthritis, lupus, multiple sclerosis, and Type 1 diabetes, the immune system's WBCs, particularly T cells and B cells, lose their ability to distinguish between self and nonself-antigens. This leads to an immune response against the body's own tissues.
    - **Monitoring WBCs:** Routine blood tests are used to monitor WBC counts and their subsets in patients with autoimmune diseases. Elevated counts, abnormal cell ratios, or the presence of autoantibodies can indicate immune dysregulation.
- **Immunosuppressive Medications**
    - **Targeting T Cells:** Many autoimmune diseases involve the overactivity of T cells, which play a central role in immune responses. Immunosuppressive drugs like corticosteroids, methotrexate, and calcineurin inhibitors are used to reduce T cell activity.
    - **B Cell Suppression:** B cells produce autoantibodies in some autoimmune diseases. Medications like rituximab specifically target B cells, reducing their numbers and antibody production.
- **Anti-Inflammatory Effects**
    - **WBC-Mediated Inflammation:** In autoimmune diseases, WBCs, especially macrophages and neutrophils, contribute to chronic inflammation by releasing pro-inflammatory cytokines and chemokines.
    - **Anti-Inflammatory Medications:** Drugs like nonsteroidal anti-inflammatory drugs (NSAIDs) and biologics, such as tumor necrosis factor (TNF) inhibitors, can reduce inflammation by acting on various WBC-mediated pathways (Zhang et al., 2015).
- **Immune Modulation**
    - **Regulatory T Cells (Tregs):** Tregs are a subset of T cells that play a crucial role in maintaining immune tolerance and preventing autoimmune responses. Therapies aimed at boosting Treg activity are being explored.
    - **Cytokine Modulation:** Cytokines produced by WBCs, like interleukin-10 (IL-10) and transforming growth factor-beta (TGF-β), can have immunosuppressive effects. Medications that promote these cytokines may help in managing autoimmune diseases (Nauseef&Borregaard, 2014).
- **Leukapheresis**
    - **Selective Removal of WBCs:** In some severe autoimmune conditions, leukapheresis, a procedure that removes WBCs from the

blood, can be considered. This approach is used to rapidly reduce the number of immune cells involved in the autoimmune response.

3.50. RESEARCH AND THERAPEUTIC ADVANCES

- **Ongoing Research:** The understanding of WBC involvement in autoimmune diseases is constantly evolving. Researchers are exploring novel therapies that target specific immune pathways, such as Janus kinase (JAK) inhibitors and interleukin inhibitors.
- **Monitoring and Management**
  - **Regular Monitoring:** Patients with autoimmune diseases require regular monitoring of their WBC counts, autoantibodies, and disease activity. This helps healthcare providers make informed decisions about treatment adjustments.

    In summary, white blood cells are central players in autoimmune diseases, both as contributors to immune dysregulation and as targets for therapy. Managing autoimmune diseases often involves modulating WBC activity and reducing inflammation to alleviate symptoms and slow disease progression. Ongoing research is focused on developing more targeted and personalized approaches to improve the management of autoimmune conditions.

3.51. EMERGING RESEARCH: PRECISION MEDICINE AND BEYOND

Emerging research in the field of precision medicine is transforming the way we understand and treat diseases, including those involving white blood cells. Here are some key insights into this exciting area of study:

- **Personalized Treatment Approaches**
  - **Genomic Profiling:** Precision medicine involves analyzing an individual's genetic makeup, including variations in genes related to the immune system. This information helps identify specific genetic factors that may contribute to diseases involving white blood cells.
  - **Tailored Therapies:** With a better understanding of a patient's genetic profile, healthcare providers can select treatments that are more likely to be effective and have fewer side effects. This approach is particularly important in autoimmune diseases and cancer, where the immune system plays a significant role (Zhang et al., 2015).
- **Immunotherapy Advancements**

- o **Immunogenomics:** Researchers are exploring how the genetic characteristics of white blood cells, such as T cells and B cells, influence their behavior. This information can guide the development of targeted immunotherapies for conditions like cancer and autoimmune diseases.
- o **CAR-T Cell Therapy:** Chimeric antigen receptor T-cell (CAR-T) therapy is a groundbreaking immunotherapy that involves genetically engineering a patient's T cells to target specific cancer cells. This approach is showing promise in treating various blood cancers.

- **Biomarker Discovery**
  - o **Biomarker Identification:** Researchers are actively searching for biomarkers—molecular indicators in the blood or tissues—that can predict disease onset, progression, or response to treatment. These biomarkers often involve specific white blood cell subsets and their activities.
  - o **Early Diagnosis:** Biomarkers associated with white blood cell function may enable earlier diagnosis of conditions like leukemia, lymphoma, and autoimmune diseases, allowing for more timely interventions.
- **Advanced Technologies**
  - o **Single-Cell Analysis:** Cutting-edge technologies like single-cell RNA sequencing are providing unprecedented insights into the heterogeneity of white blood cell populations. This technology allows researchers to study individual cells and their gene expression profiles.
  - o **CRISPR-Cas9:** Gene-editing techniques, such as CRISPR-Cas9, are being explored to modify white blood cells for therapeutic purposes. This approach has the potential to correct genetic defects or enhance immune responses. For example, Bhokisham et al. (2023) presented an approach to using CRISPR-Cas9 engineering in autologous CAR-T therapy, which is shown in Figure 3.32.
- **Drug Development**
  - o **Targeted Therapies:** Precision medicine is driving the development of targeted therapies that focus on specific molecules or pathways within white blood cells. These therapies aim to modulate immune responses with greater precision.
  - o **Combination Therapies:**Researchers are investigating combinations of drugs that target different aspects of white blood cell function. This approach may enhance treatment efficacy and reduce the risk of resistance.

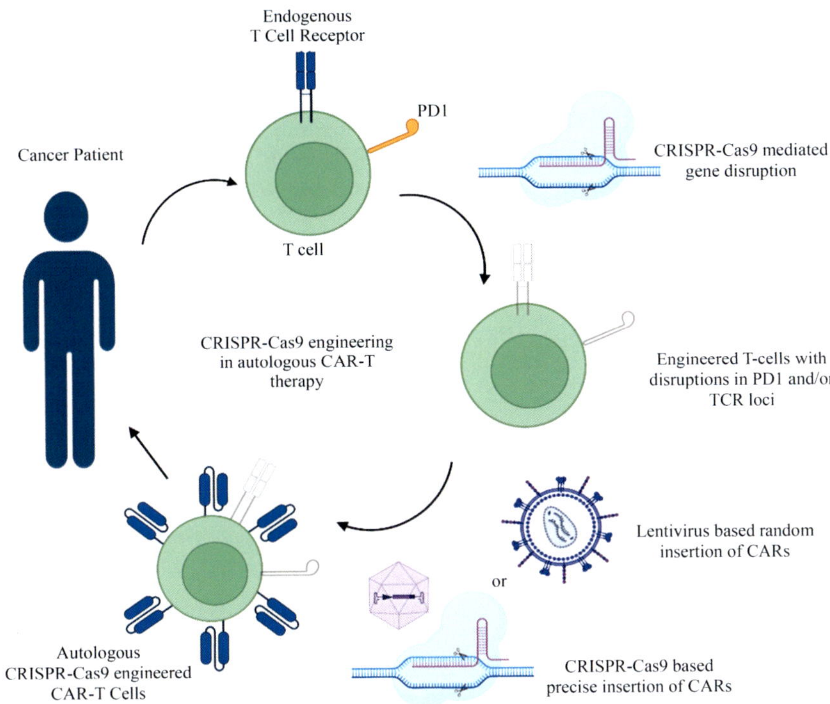

Figure 3.32.*CRISPR-Cas9 engineering in autologous CAR-T therapy.*

Source: Bhokisham N, Laudermilch E, Traeger LL, Bonilla TD, Ruiz-Estevez M, & Becker JR, (2023). CRISPR-Cas system: The current and emerging translational landscape. Cells, 12(8), 1103. https://doi.org/10.3390/cells12081103: © 2023 by the authors. Licensee MDPI, Basel, Switzerland. This article is an open access article distributed under the terms and conditions of the Creative Commons Attribution (CC BY) license (https://creativecommons.org/licenses/by/4.0/).

- **Data Integration**
    - o **Big Data Analysis:** Precision medicine relies on the integration of diverse data sources, including genomics, proteomics, and clinical data. Advanced data analytics and artificial intelligence (AI) are helping researchers identify patterns and correlations that inform treatment decisions.
- **Patient-Centric Care**
    - o **Shared Decision-Making:** Precision medicine promotes shared decision-making between patients and healthcare providers. Patients are increasingly empowered to participate in treatment decisions based on their unique genetic and health profiles.

Emerging research in precision medicine holds tremendous promise for improving the diagnosis and treatment of diseases involving white blood cells. By tailoring therapies to individual patients and gaining a deeper understanding of the genetic and molecular factors at play, researchers aim to enhance treatment outcomes and minimize adverse effects. This approach represents a shift towards more personalized and effective healthcare.

- **Artificial Intelligence and WBC Analysis:** Artificial intelligence (AI) is revolutionizing many aspects of healthcare, including the analysis of white blood cells (WBCs). Here's an explanation of how AI is being used in WBC analysis (Zhang et al., 2015):
- **Disease Diagnosis and Classification**
    - **Automated Detection of Abnormalities:** AI algorithms can analyze blood smears and other microscopy images to identify and classify white blood cell abnormalities. This includes detecting conditions like leukemia, where there is an abnormal increase in immature white blood cells (blast cells). AI can help pathologists by automating the screening process and providing faster and more consistent results.
    - **Subtype Classification:** AI can differentiate between different subtypes of white blood cells, such as neutrophils, lymphocytes, eosinophils, and monocytes. This is crucial for diagnosing various infections, autoimmune diseases, and blood disorders.
- **Early Disease Detection**
    - **Predictive Analytics:** AI models can analyze patient data, including WBC counts and trends, to predict the likelihood of developing certain diseases or conditions. For example, changes in WBC counts over time may indicate the early stages of an infection or an inflammatory disorder.

      As an example, the use of a neural network to classify white blood cells by Cheuque et al. (2022) is given in Figure 3.33. The tool was applied to classify lymphocytes and monocytes (mononuclear) and segmented neutrophils and eosinophils, and it recorded an accuracy of 98.4% (accuracy, recall, precision, and F1-score). Such findings and further research can pave the way for the use of AI tools in assisting a pathologist in analyzing blood smears.
    - **Identifying Rare Conditions:** Some rare diseases and genetic disorders can affect white blood cell morphology. AI can assist in identifying these rare conditions by recognizing unique patterns in blood samples.

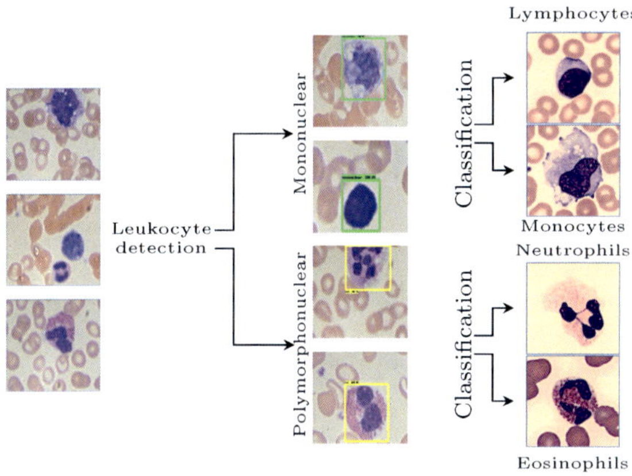

Figure 3.33.*Neural network to classify white blood cells.*

Source: Cheuque C, Querales M, León R, Salas R, & Torres R, (2022). An efficient multi-level Convolutional neural network approach for white blood cells classification. Diagnostics, 12(2), 248. https://doi.org/10.3390/diagnostics12020248.

- **Treatment Monitoring**
  - **Response to Therapy:** AI can track changes in WBC counts and cell morphology during treatment. This helps healthcare providers assess how well a patient is responding to therapy. For example, in cancer treatment, monitoring WBC changes can indicate the effectiveness of chemotherapy.
  - **Side Effect Prediction:** AI algorithms can predict potential side effects of medications, including those affecting white blood cell production. This enables healthcare providers to make informed decisions about treatment options.
- **Speed and Efficiency**
  - **Faster Analysis:** AI can process large volumes of WBC data rapidly. This is particularly useful in emergency situations where quick decisions about treatment are crucial.
  - **Reduced Workload:** By automating routine WBC analysis tasks, AI can reduce the workload on healthcare professionals, allowing them to focus on more complex aspects of patient care.
- **Research and Drug Development**
  - **Data Mining:** AI can mine vast amounts of medical literature and research data to identify trends, associations, and potential drug targets related to white blood cell function and diseases. This aids in drug discovery and development.

As an example, a study by Davis et al (2023) is presented in Figure 3.34. Given the ongoing quest for biomarkers to detect cancer types, the researchers used flow cytometry to identify blood leukocyte subset signatures. These could be used to train machine learning models to predict both the presence and type of cancer in their animal models.

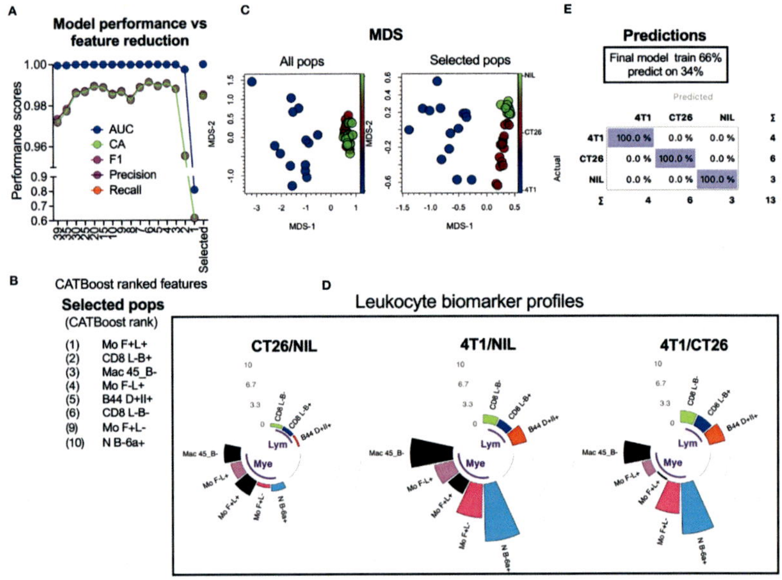

Figure 3.34. Flow cytometry to identify blood leukocyte subset signatures

Source: Simon Davis DA, Ritchie M, Hammill D, Garrett J, Slater RO, Otoo N, Orlov A, Gosling K, Price J, Yip D, Jung K, Syed FM, Atmosukarto II and Quah BJC, (2023) Identifying cancer-associated leukocyte profiles using high-resolution flow cytometry screening and machine learning. Front. Immunol., 14, 1211064. doi: 10.3389/fimmu.2023.1211064:

3.52. PERSONALIZED MEDICINE

- **Treatment Recommendations***:* AI can analyze a patient's complete medical history, including WBC-related data, to recommend personalized treatment plans. This is especially important in conditions where WBCs play a central role, such as autoimmune diseases.
- **Telemedicine and Remote Monitoring**
  - **Remote WBC Analysis:** AI-powered mobile apps and devices can enable patients to monitor their WBC counts at home. The data can be transmitted to healthcare providers for real-time analysis and intervention.
- Challenges and Considerations
  - **Data Quality:** AI models require high-quality data for training and validation. Ensuring the accuracy and reliability of WBC data is crucial.
  - **Ethical and Privacy Concerns:** Protecting patient privacy and ensuring the ethical use of AI in healthcare is a paramount concern.
  - **Regulatory Approval**: AI-based medical devices and software must go through rigorous regulatory processes to ensure safety and efficacy (Zhang et al., 2015).
  - **Interdisciplinary Collaboration:** Effective AI implementation in healthcare often requires collaboration between computer scientists, clinicians, and pathologists.

AI is poised to transform the field of WBC analysis by improving accuracy, speed, and the ability to detect subtle patterns and trends in large datasets. As technology advances and AI models become more sophisticated, their role in healthcare, particularly in the analysis of white blood cells, is likely to continue expanding.

- **Nanotechnology and Drug Delivery:** White blood cells (WBCs), also known as leukocytes, have shown promising applications in nanotechnology and drug delivery. Here's how they are being utilized in these fields:
- **Targeted Drug Delivery**
  - **Nanoparticle Carriers:** Researchers have developed nanoparticles that can be loaded with therapeutic drugs. These nanoparticles can then be coated with a layer of WBC membranes or functionalized with WBC-specific ligands. The resulting nanostructures, often referred to as "leukosomes" or "WBC-mimicking nanoparticles,"

can carry drugs directly to sites of infection, inflammation, or tumors. The surface modifications allow these nanoparticles to evade the immune system and enhance their specificity for target tissues.

Vijayan et al. (2018) proposed a scheme for the use of specific nanoparticles for treatment, which is presented in Figure 3.35.

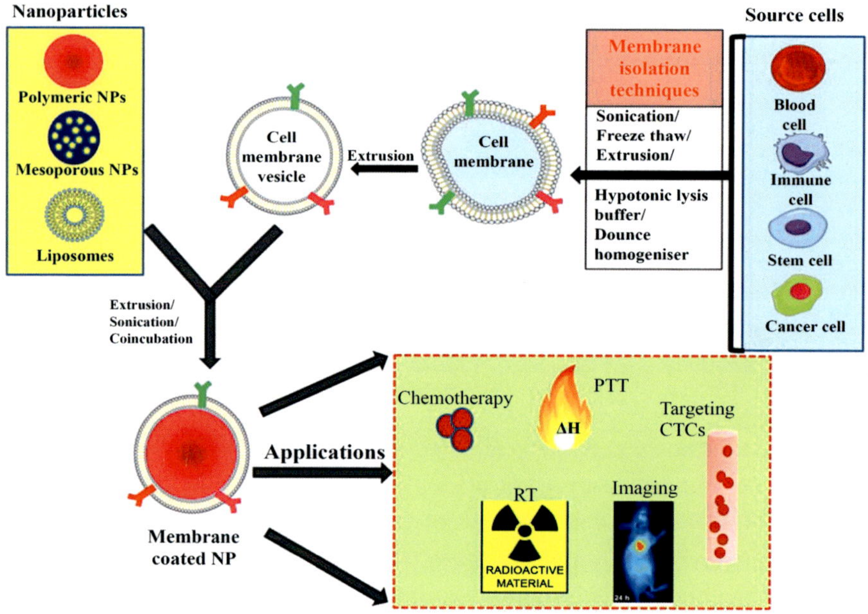

Figure 3.35. *A scheme for the use of specific nanoparticles for treatment.*

Key: TT—photothermal therapy, CTC—circulating tumor cells, RT—radiotherapy, NP—nanoparticle

Source: Vijayan V, Uthaman S, & Park I-K, (2018). Cell membrane-camouflaged nanoparticles: A promising biomimetic strategy for cancer theragnostics. Polymers, 10(9), 983. https://doi.org/10.3390/polym10090983: © 2018 by the authors. Licensee MDPI, Basel, Switzerland. This article is an open access article distributed under the terms and conditions of the Creative Commons Attribution (CC BY) license (http://creativecommons.org/licenses/by/4.0/).

- **Improved Drug Distribution:** WBC-based drug carriers can improve drug distribution to hard-to-reach areas, such as solid tumors or sites of infection. This targeted delivery reduces the exposure of healthy tissues to the drug, minimizing side effects.
- **Inflammatory Disease Treatment:**In conditions like rheumatoid arthritis, where chronic inflammation is a hallmark, WBC-based drug

carriers can specifically deliver anti-inflammatory agents to affected joints, improving therapeutic outcomes.

- **Immune System Modulation**
  - o **Immunotherapy:** WBCs can be engineered to carry therapeutic agents or antibodies that modulate the immune response. For example, macrophages can be loaded with immunomodulatory drugs and redirected to sites of inflammation or autoimmune reactions.
  - o **Cancer Immunotherapy:** WBCs, such as dendritic cells, can be used to present tumor-specific antigens to the immune system, activating an anti-cancer immune response. This approach is being explored in cancer immunotherapies like dendritic cell vaccines.

3.53. DIAGNOSTIC AND IMAGING APPLICATIONS

- **Contrast Agents:** WBCs can be labeled with imaging agents, such as nanoparticles or contrast dyes, for use in medical imaging techniques like magnetic resonance imaging (MRI) or positron emission tomography (PET). This allows for the visualization of inflammation, infection, or specific cell populations *in vivo*.
- **Immune System Monitoring**
  - o **Biosensors:** Researchers are developing biosensors that use WBCs or their components to detect specific biomarkers associated with diseases. These biosensors can provide real-time monitoring of a patient's immune response and disease progression.
- **Challenges and Considerations**
  - o **Biocompatibility:** Ensuring that the interaction between synthetic nanoparticles and WBCs is biocompatible and does not trigger undesirable immune responses is crucial.
  - o **Safety:** The use of WBC-based therapies must be rigorously tested for safety, as any unintended side effects could have significant consequences.
  - o **Efficiency:** Achieving high drug-loading efficiencies in WBC-based drug carriers is an ongoing challenge.
  - o **Clinical Translation:** Moving from preclinical research to clinical trials and eventual patient use requires navigating regulatory processes and addressing scalability and cost considerations.

WBCs are a versatile and promising tool in nanotechnology and drug delivery due to their natural targeting abilities and interactions with the immune system. As research in this area continues to advance, it holds the potential to revolutionize the way we treat diseases and deliver therapies.

3.54. CONCLUSION: A FUTURE AMPLIFIED TECHNOLOGY

The concluding section, "A Future Amplified Technology for White Blood Cells," likely refers to the potential advancements and technologies that can enhance our understanding and applications of white blood cells (WBCs) in various fields. Here's an explanation of what this section might entail:

- **Advanced Diagnostic Tools:** In the future, we can expect the development of more advanced diagnostic tools that leverage cutting-edge technologies like artificial intelligence and nanotechnology to detect diseases and conditions based on WBC signatures. These tools may offer quicker and more accurate diagnoses, leading to better patient outcomes.
- **Personalized Medicine:** As our understanding of WBC behavior and their role in health and disease deepens, we can anticipate the growth of personalized medicine approaches. Tailoring treatments to an individual's unique immune profile, which includes WBC activity, can optimize therapeutic interventions and minimize side effects.
- **Immunotherapies:** Immunotherapies that use engineered WBCs, such as CAR-T cell therapy, will continue to evolve. These therapies have shown remarkable success in treating certain cancers and autoimmune diseases, and ongoing research may expand their applications.
- **Drug Delivery and Targeting:** The use of WBC-based drug delivery systems will likely become more refined and widespread. These systems can improve drug efficacy while minimizing side effects, leading to more effective treatments for a variety of conditions.
- **Disease Monitoring:** Continuous monitoring of an individual's immune status, achieved through wearable devices or implantable sensors, could become a reality. This would enable real-time tracking of immune responses, allowing for early detection of infections or disease relapses.
- **Enhanced Vaccines:** Advancements in WBC research may lead to the development of more effective vaccines. Understanding how different types of white blood cells respond to pathogens can inform vaccine design, resulting in improved protection against infectious diseases.
- **Gene Editing and Therapeutics:** Technologies like CRISPR-Cas9

may be used to modify WBCs for therapeutic purposes. Precise gene editing could correct genetic immune disorders or enhance immune responses to fight diseases.

- **AI-Driven Discoveries:** Artificial intelligence and machine learning will play a significant role in analyzing vast datasets related to WBC behavior and immune responses. AI algorithms can identify patterns and correlations that human researchers might miss, leading to new insights and breakthroughs.

3.55. ETHICAL CONSIDERATIONS

With these technological advancements, ethical considerations regarding genetic manipulation, patient consent, and data privacy will become even more critical. Balancing the potential benefits of these technologies with their ethical implications will be an ongoing challenge. Thus, the future holds tremendous promise for WBC-related technologies across various domains of medicine and biotechnology. These advancements have the potential to revolutionize healthcare, offering more precise diagnostics, tailored treatments, and improved disease management. However, they also come with ethical and regulatory responsibilities that must be carefully addressed as these technologies continue to evolve.

"Research is seeing what everybody else has seen and thinking what nobody else has thought."

—Albert Szent-Györgyi

REFERENCES

1. Adrover, J. M., Nicolás-Ávila, J. A., & Hidalgo, A. (2016). Aging: A temporal dimension for neutrophils. *Trends in Immunology, 37*(5), 334–345. https://doi.org/10.1016/j.it.2016.03.005.
2. Ahmed, A., Khan, Q., Naeem, M., Iqbal, M., Anpalagan, A., &Awais, M. (2018). An insight to the performance of estimation of distribution algorithm for multiple line outage identification. *Swarm and Evolutionary Computation, 39*, 114–122. https://doi.org/10.1016/j.swevo.2017.09.006.
3. Azad, A. K., Torrelles, J. B., & Schlesinger, L. S. (2008). Mutation in the DC-SIGN cytoplasmic triacidic cluster motif markedly attenuates receptor activity for phagocytosis and endocytosis of mannose-containing ligands by human myeloid cells. *Journal of Leukocyte Biology, 84*(6), 1594–1603. https://doi.org/10.1189/jlb.0308192.
4. Bakema, J. E., & van Egmond, M. (2011). The human immunoglobulin A Fc receptor FcαRI: A multifaceted regulator of mucosal immunity. *Mucosal Immunology, 4*(6), 612–624. https://doi.org/10.1038/mi.2011.36.
5. Busso, N., & So, A. (2012). Microcrystals as DAMPs and their role in joint inflammation. *Rheumatology, 51*(7), 1154–1160. https://doi.org/10.1093/rheumatology/ker524.
6. Canton, J. (2014). Phagosome maturation in polarized macrophages. *Journal of Leukocyte Biology, 96*(5), 729–738. https://doi.org/10.1189/jlb.1MR0114-021R.
7. Chen, F., Wu, W., Millman, A., Craft, J. F., Chen, E., Patel, N., et al. (2014). Neutrophils prime a long-lived effector macrophage phenotype that mediates accelerated helminth expulsion. *Nature Immunology, 15*(10), 938–946. https://doi.org/10.1038/ni.2984.
8. Dambuza, I. M., & Brown, G. D. (2015). C-type lectins in immunity: Recent developments. *Current Opinion in Immunology, 32*, 21–27. https://doi.org/10.1016/j.coi.2014.12.002.
9. Eash, K. J., Greenbaum, A. M., Gopalan, P. K., & Link, D. C. (2010). CXCR2 and CXCR4 antagonistically regulate neutrophil trafficking from murine bone marrow. *Journal of Clinical Investigation, 120*(7), 2423–2431. https://doi.org/10.1172/JCI41649.
10. Ericson, J. A., Duffau, P., Yasuda, K., Ortiz-Lopez, A., Rothamel, K., Rifkin, I. R., et al. (2014). Gene expression during the generation and activation of mouse neutrophils: Implication of novel functional and regulatory pathways. *PLoS ONE, 9*(10), e108553. https://doi.org/10.1371/journal.pone.0108553.

11. Fuchs, T. A., Abed, U., Goosmann, C., Hurwitz, R., Schulze, I., Wahn, V., et al. (2007). Novel cell death program leads to neutrophil extracellular traps. *Journal of Cell Biology, 176*(2), 231–241. https://doi.org/10.1083/jcb.200606027.
12. Garg, A. D., Dudek-Peric, A. M., Romano, E., &Agostinis, P. (2015a). Immunogenic cell death. *International Journal of Developmental Biology, 59*, 131–140. https://doi.org/10.1387/ijdb.150061pa.
13. Garg, A. D., Galluzzi, L., Apetoh, L., Baert, T., Birge, R. B., Bravo-San Pedro, J. M., et al. (2015b). Molecular and translational classifications of DAMPs in immunogenic cell death. *Frontiers in Immunology, 6*, Article 588. https://doi.org/10.3389/fimmu.2015.00588.
14. Giepmans, B. N., Adams, S. R., Ellisman, M. H., &Tsien, R. Y. (2006). The fluorescent toolbox for assessing protein location and function. *Science, 312*(5771), 217–224.
15. Gordon, S. (2016). Phagocytosis: An immunobiologic process. *Immunity, 44*(3), 463–475. https://doi.org/10.1016/j.immuni.2016.02.026.
16. Greenlee-Wacker, M. C. (2016). Clearance of apoptotic neutrophils and resolution of inflammation. *Immunological Reviews, 273*(1), 357–370. https://doi.org/10.1111/imr.12453.
17. Levin, R., Grinstein, S., &Schlam, D. (2015). Phosphoinositides in phagocytosis and micropinocytosis. *Biochimica et Biophysica Acta (BBA) - Molecular and Cell Biology of Lipids, 1851*(6), 805–823. https://doi.org/10.1016/j.bbalip.2014.09.005.
18. Lippincott-Schwartz, J. (2004). Dynamics of secretory membrane trafficking. *Annals of the New York Academy of Sciences, 1038*, 115–124.
19. Martin, C., Burdon, P. C., Bridger, G., Gutierrez-Ramos, J. C., Williams, T. J., & Rankin, S. M. (2003). Chemokines acting via CXCR2 and CXCR4 control the release of neutrophils from the bone marrow and their return following senescence. *Immunity, 19*(4), 583–593. https://doi.org/10.1016/S1074-7613(03)00263-2.
20. Nauseef, W. M., &Borregaard, N. (2014). Neutrophils at work. *Nature Immunology, 15*(6), 602–611. https://doi.org/10.1038/ni.2921.
21. Netea, M. G., Joosten, L. A., Latz, E., Mills, K. H., Natoli, G., Stunnenberg, H. G., et al. (2016). Trained immunity: A program of innate immune memory in health and disease. *Science, 352*(6284), aaf1098. https://doi.org/10.1126/science.aaf1098.
22. Pushparaj, P. N., Tay, H. K., Wang, C. C., Hong, W., & Melendez, A. J. (2009). VAMP8 is essential in anaphylatoxin-induced degranulation,

TNF-α secretion, peritonitis, and systemic inflammation. *Journal of Immunology, 183*(2), 1413–1418.

23. Reefman, E., Kay, J. G., Wood, S. M., et al. (2010). Cytokine secretion is distinct from secretion of cytotoxic granules in NK cells. *Journal of Immunology, 184*(9), 4852–4862.
24. Rosales, C., & Uribe-Querol, E. (2013). Fc receptors: Cell activators of antibody functions. *Advances in Bioscience and Biotechnology, 4*, 21–33. https://doi.org/10.4236/abb.2013.44A004.
25. Scapini, P., &Cassatella, M. A. (2014). Social networking of human neutrophils within the immune system. *Blood, 124*(5), 710–719. https://doi.org/10.1182/blood-2014-03-453217.
26. Summers, C., Rankin, S. M., Condliffe, A. M., Singh, N., Peters, A. M., &Chilvers, E. R. (2010). Neutrophil kinetics in health and disease. *Trends in Immunology, 31*(8), 318–324. https://doi.org/10.1016/j.it.2010.05.006.
27. Tecchio, C., Micheletti, A., &Cassatella, M. A. (2014). Neutrophil-derived cytokines: Facts beyond expression. *Frontiers in Immunology, 5*, Article 508. https://doi.org/10.3389/fimmu.2014.00508.
28. Woodfin, A., Voisin, M. B., Beyrau, M., Colom, B., Caille, D., Diapouli, F. M., et al. (2011). The junctional adhesion molecule JAM-C regulates polarized trans endothelial migration of neutrophils *in vivo*. *Nature Immunology, 12*(8), 761–769. https://doi.org/10.1038/ni.2062.
29. Zhang, D., Chen, G., Manwani, D., Mortha, A., Xu, C., Faith, J. J., et al. (2015). Neutrophil ageing is regulated by the microbiome. *Nature, 525*(7569), 528–532. https://doi.org/10.1038/nature15367.
30. Zhang, P., Schwartz, O., Pantelic, M., Li, G., Knazze, Q., Nobile, C., et al. (2006). DC-SIGN (CD209) recognition of Neisseria gonorrhoeae is circumvented by lipooligosaccharide variation. *Journal of Leukocyte Biology, 79*(4), 731–738. https://doi.org/10.1189/jlb.0405184.

Chapter 4

BLOOD DONORS AND DONATION AND MANUFACTURING BLOOD: A RESEARCH PERSPECTIVE

"With every heart beat we live to see the sunset, with each drop of blood donated; another precious life will live to see the sunrise."

—*Wayne Chirisa*

4.1. BLOOD DONATION: WHY THE HESITATION-RESEARCH OFFERS CLUES?

Given that blood transfusions serve as an integral component of healthcare, a 2019-published study funded by the National Institutes of Health scrutinized the blood transfusion need and supply at national levels. The blood availability for 180 (92%) of 195 countries from the WHO Global Status Report on Blood Safety and Availability was calculated from the blood component preparation data of 2011–13. The disease-specific transfusion needs per prevalent case for 20 causes in the USA were computed with the National (Nationwide) Inpatient Sample dataset (2000 and 2014) and the State Inpatient Databases from the Healthcare Cost and Utilization Project (2003–2007). The results of the Global Burden of Disease (GBD) 2017 study were applied to derive the unmet need as the difference between the estimated supply and need.

The researchers stated that the gap was large, especially for low-income and middle-income countries. While the global blood supply was 272,270,243 (268,002,639–276,698,494) blood product units, the global blood need was 304,711,244 (95% uncertainty interval [UI] 293,064,637–314,049,479) in 2017 with the need-to-supply ratio at 1.12 (95% UI 1.07–1.16). Shockingly,

61% (119/195) of the countries had an unmet need of 102,359,632 (95% UI 93,381,710–111,360,725) blood product units, which can be translated to 1,849 (1,687–2,011) units per 100,000 population globally. This hence, highlighted the critical importance of national transfusion services and blood transfusion (Roberts, James, Delaney, & Fitzmaurice, 2019).

Given this serious scenario, let us examine a few studies reported to try to understand the associated factors. Research has outlined the motivators of blood donation to be extrinsic rewards like compensation and altruism, while the deterrents were mainly fear, including the fear of pain, the puncture, or the flow of blood through the collection bag (Li, Lei, Li, Zhao, Dai, Jin, et al., 2021).

A systematic meta-analysis of antecedents of blood donation behavior and intentions by Bednall et al. (2013) analyzed 24 predictive correlational studies of donation behavior and 37 studies of donation intentions. The following programs were taken into account: "(1) the Theory of Planned Behavior (TPB), (2) prosocial motivation, (3) affective expectations, (4) donor site experience, (5) past donation behavior, and (6) donor demographics." Positive associations emerged for the TPB-"intentions to donate, perceived behavioral control, attitude toward donation, self-efficacy, and donor role identity." An anticipated regret for not donating, past donations, and donor age were other positively associated factors. A negative association was seen for the experiences of a donor at a collection site. The researchers summarized the necessity of minimizing the risk of adverse reactions (as seen with experiences during collection) and taking into account their attitudes to boost donation (Bednall et al., 2013).

A 2021-published study by Li et al. (2021) aimed to comprehend the intention to donate again from a social cognitive perspective. The researchers employed a cross-sectional survey that was conducted in six cities in Jiangsu, China that were geographically and socioeconomically variable. The approach entailed a self-administered questionnaire conducted by two well-trained medical students. Subsequent analyses of 191 blood donors revealed that donors with higher fear scores had a lower intention to return ($p = 0.008$). While self-efficacy and intention to return showed a statistically significant positive association ($p < 0.001$), there was no significance between the association among intrinsic rewards ($p = 0.387$), extrinsic rewards ($p = 0.939$), and intention to return.

The tables below as reported by the researchers have been presented below for visual comprehension (*Copyright © 2021 Li, Lei, Li, Zhao, Dai, Jin, Fu, Cai, Lin and Tu. This is an open-access article distributed under the terms of the Creative Commons Attribution License (CC BY). The use, distribution or reproduction in other forums is permitted, provided the original author(s) and the copyright owner(s) are credited and that the original publication in this journal is cited, in accordance with accepted academic practice. No use, distribution or reproduction is permitted which does not comply with these terms*).

Table 4.1. Higher fear deters blood donation

Variables	Overall *N* (column %)	Intention score median (p25,p75)	*p*-value
Gender			0.976
Female	88 (46.1)	4 (3.67, 4.67)	
Male	103 (53.9)	4 (3.67, 4.67)	
Age	28 (23, 38)		0.204
18–24	66 (34.7)	4 (3.33, 4.33)	
25–34	62 (32.6)	4 (3.33, 4.67)	
35–57	62 (32.6)	4 (3.67, 4.67)	
Monthly income*	3,000 (1,600, 4,000)	NA	NA
Place of residence			0.697
Urban	140 (74.1)	4 (3.67, 4.67)	
Rural	49 (25.9)	4 (3.67, 4.67)	
Place of birth			0.365
Urban	82 (44.1)	4 (3.67, 4.67)	
Rural	104 (55.9)	4 (3.67, 4.67)	
Marital status			0.042
Married	106 (56.7)	4 (3.67, 4.67)	
Unmarried	81 (43.3)	4 (3.33, 4.33)	
Educational attainment			0.172
College or above	103 (56.0)	4 (3.33, 4.33)	
High school or less	81 (44.0)	4 (3.67, 4.67)	
Donation history			0.003
Repeated donors	105 (55.0)	4 (3.33, 4.33)	
First-time donors	86 (45.0)	4 (3.67, 4.67)	

**In Chinese Yuan. p is the probability value for the comparison of intention scores by different sociodemographic groups. Because the variables of age [result of Shapiro–Wilk test: p (skewness) < 0.001, p (Kurtosis) = 0.010, p (joint) < 0.001], monthly income [result of Shapiro-Wilk test: p (skewness) < 0.001, p (Kurtosis) < 0.001, p (joint) < 0.001], and intention score are not distributed normally, median (P25, P75) was used to describe intention scores.*

Variables	Coefficients (95% CI)	*p*-value	Weights of individual predictors
Marital status			4.0%
Married (Ref)			
Unmarried	−0.076 (−0.294, 0.142)	0.495	
Donation history			4.6%
First-time donors (Ref)			
Repeated donors	0.180 (−0.032, 0.392)	0.096	
Monthly income	0 (0, 0)	0.238	2.1%
Fear	−0.150 (−0.261, −0.039)	0.008	9.5%
Intrinsic rewards	0.097 (−0.123, 0.318)	0.387	23.5%
Extrinsic rewards	−0.005 (−0.126, 0.116)	0.939	1.3%
Self-efficacy	0.542 (0.319, 0.765)	<0.001	33.6%
Constant	1.704 (0.814, 2.593)	<0.001	

Ref, reference group. Log likelihood = −115.45, AIC = 1.76, and BIC = −609.65. Weight of individual predictors were calculated with dominance analysis after the estimation of generalized linear model (overall fit statistic = 0.413).

This study highlighted the necessity to focus more on addressing the fear aspect rather than the "rewards."

An interesting research article by Renaux and team (2022) assessed the efficacy of posters in waiting rooms of general practitioners (GPs) given that such waiting rooms can serve as potentially valuable sites for spreading necessary education. The study was an observational, multicenter, prospective study at six blood donation centers in the Aube Department of France from 1 June to 31 December 2021 where posters inviting people to donate blood were displayed. Of the 1,469 questionnaires that were completed by blood donors: 729 reported having seen the posters, and 740 reported not having seen the posters. Further, 68% of all respondents indicated that GP room posters motivated them to donate blood more often. Many claimed that the donation was due to online publicity (7.5 vs. 3.9%, adjusted Odds ratio [aOR] 1.75, 95% confidence interval [CI] 1.12–2.82, p = 0.02) and television advertisements (8.0 vs. 4.2%, aOR 1.74, 95%CI 1.10–2.76, p = 0.02). The outcomes as compiled by the researchers have been presented below.

Table 4.2 A study investigating the efficacy of blood donation posters in waiting rooms of general practitioners

	Overall *N* = 1,469	**Saw posters** *N* = 729	**Did not see posters** *N* = 740	***P*-value**
Age, years	45.3 ± 14.7	45.0 ± 15.1	45.5 ± 14.4	0.55
Sex				0.95
Female	761 (51.9)	375 (51.8)	386 (51.9)	
Male	706 (48.1)	349 (48.2)	357 (48.1)	
Not a first-time donor	1345 (91.7)	655 (90.5)	690 (92.9)	0.10
Reason for attending blood drive				
Accompanied/encouraged by a close friend or relative	132 (9.0)	72 (9.9)	60 (8.1)	0.21
Prompted by EFS television ads	89 (6.1)	58 (8.0)	3 (4.2)	0.002
Prompted by EFS advertisement on the internet	83 (5.7)	54 (7.5)	29 (3.9)	0.003
In response to an invitation sent by the EFS by text message	593 (40.4)	294 (40.7)	299 (40.2)	0.87
In response to an invitation sent by the EFS by e-mail	263 (17.9)	124 (17.1)	139 (18.7)	0.43
A friend or member of my family has a disease that requires blood transfusion	134 (9.1)	58 (8.0)	76 (10.3)	0.14
Opportunity: The blood drive is being held at a time when I'm free	709 (48.3)	354 (48.9)	355 (47.8)	0.67
Altruism: It's important to help people who are ill	1,097 (74.8)	549 (75.8)	548 (73.8)	0.36
I'm interested by the food provided after the blood donation	42 (2.9)	23 (3.2)	19 (2.6)	0.47
I'm interested by the blood test / health check-up provided	97 (6.6)	50 (6.9)	47 (6.3)	0.65
GP participating in the study	518 (35.3)	272 (37.6)	246 (33.1)	0.08
I believe that posters in the GP's waiting room would prompt me to give blood more often.	997 (68.0)	521 (72.1)	476 (64.1)	0.001

EFS, French Blood Transfusion Service, GP, general practitioner.

	Donors who saw the posters in a GP's waiting room $N = 729$	Donors who did not see the posters in a GP's waiting room $N = 740$	OR (95% CI)	*P*-value
Motivation for participation				
Saw a television advertisement	58 (8.0)	3 (4.2)	1.74 (1.10–2.76)	0.02
Saw an advertisement on internet	54 (7.5)	29 (3.9)	1.75 (1.12–2.82)	0.02
GP participating in study	272 (37.6)	246 (33.1)	1.22 (0.98–1.51)	0.08

GP, general practitioner; OR, odds ratio; CI, confidence interval.

Source: (Renaux, Bouazzi, Sanchez, Hottois, Martin, Chrusciel, & Sanchez, 2022) Impact of promoting blood donation in general practice: Prospective study among blood donors in France. Front. Public Health 10: 1080096. doi: 10.3389/fpubh.2022.1080096:

A 2021-published work by Greffin and team qualitatively probed the personal motives and deterrents associated with blood donation in a German population cohort. The study entailed a face-to-face cross-sectional survey of 2531 respondents which was "representative of the German population in terms of age, gender, and residency." Interestingly, two-thirds of the population shared that they had never donated blood before. The motives and deterrents were grouped into 10 domains with 50 main categories and 65 sub-categories. The most common motives for blood donation were "altruism," "social responsibility," and "charity," while the common deterrents included "health status," "age," and "lack of time." Table 4.3 compiled by the author with the overall findings has been given for comprehension (Greffin et al., 2021).

Table 4.3. Self-reported motivators and deterrents for retrospective (lifetime) blood donation behavior

(a)								
Domain (D)/Main Category	**Subcategory**	**Non-Donors (Previous Lifetime)**		**Deferred Donors (Previous Lifetime)**		**Donors (Previous Lifetime)**		**Effect Size (Sig.)**
D01 Ineligibility		n	%	n	%	n	%	Cramer's V (p-Value)

Unspecific reason		12	0.8	3	2.3	0	0.0	0.071 (0.002)
• Specific reason	Age	110	7.3	5	3.8	7	0.8	0.144 (0.000)
	Health	207	13.8	75	56.4	19	2.1	0.365 (0.000)
	Pregnancy	6	0.4	0	0.0	0	0.0	0.040 (0.128)
• Specific reason	Other	12	0.8	8	6.0	3	0.3	0.129 (0.000)
D02 Impact/Effect	Total n/ domain (D01)	336	22.4	90	67.7	27	3.0	0.387 (0.000)
• Physical consequences	Positive	0	0.0	1	0.8	19	2.1	0.113 (0.000)
	Negative—health	67	4.5	1	0.8	4	0.4	0.118 (0.000)
	Negative—risk	36	2.4	0	0.0	1	0.1	0.094 (0.000)
• Mental/psychological wellbeing	Positive	1	0.1	0	0.0	16	1.8	0.101 (0.000)
	Negative	7	0.5	0	0.0	0	0.0	0.044 (0.091)
D03 Fears/Aversion	Total n/ domain (D02)	105	7.0	2	1.6	37	4.1	0.072 (0.001)
• Fears	Fears in general	43	2.9	0	0.0	0	0.0	0.109 (0.000)
	Fear of the needle	90	6.0	0	0.0	0	0.0	0.159 (0.000)
	Fear of blood	5	0.3	0	0.0	0	0.0	0.037 (0.180)
	Fear to donate	14	0.9	0	0.0	0	0.0	0.062 (0.008)
	Fear of pain	3	0.2	0	0.0	0	0.0	0.029 (0.358)
	Other	9	0.6	1	0.8	0	0.0	0.047 (0.062)
• Aversion	Needles	28	1.9	0	0.0	0	0.0	0.088 (0.000)
	Blood	35	2.3	0	0.0	0	0.0	0.098 (0.000)
	Other	23	1.5	0	0.0	0	0.0	0.079 (0.000)
D04 Obstacles/Barriers	Total n/ domain (D03)	219	14.6	1	0.8	0	0.0	0.253 (0.000)
Lack of information		32	2.1	1	0.8	1	0.1	0.084 (0.000)

No opportunity/Lack of possibilities		94	6.3	0	0.0	3	0.3	0.153 (0.000)
Organization/effort		15	1.0	1	0.8	0	0.0	0.059 (0.011)
Time/Lack of time		137	9.1	2	1.5	4	0.4	0.182 (0.000)
Personal reasons		46	3.1	0	0.0	1	0.1	0.108 (0.000)
D05 Norms	Total n/ domain (D04)	292	19.5	3	2.3	8	0.9	0.278 (0.000)
• Reciprocity	General (if no other category)	1	0.1	1	0.8	6	0.7	0.054 (0.026)
	Future-orientated	1	0.1	3	2.3	51	5.7	0.182 (0.000)
	Past-orientated (self)	1	0.1	2	1.5	10	1.1	0.077 (0.001)
	Past-orientated (friends and family)	2	0.2	0	0.0	5	0.5	0.040 (0.130)
Altruism		16	1.1	18	13.5	405	45.4	0.551 (0.000)
• Obligation/self-evident/social conscientiousness and responsibility		4	0.3	0	0.0	109	12.2	0.277 (0.000)
Religious reasons		10	0.7	0	0.0	3	0.3	0.028 (0.382)
Personal belief		0	0.0	1	0.8	27	3.0	0.136 (0.000)
• Important/necessary/meaningful/good cause		7	0.5	3	2.3	84	9.4	0.223 (0.000)
Vocational affiliation		1	0.1	1	0.8	28	3.1	0.134 (0.000)
Surrendering responsibility		15	1.0	0	0.0	0	0.0	0.064 (0.006)
	Total n/ domain (D05)	57	3.8	26	19.5	633	70.9	0.702 (0.000)
D06Image/Experience		N	%	n	%	n	%	Cramer's V (p-Value)
• Compensate for lack of blood products/Support the health care system		2	0.1	3	2.3	64	7.2	0.203 (0.000)
• Lack of trust (rip-off/crime/fraud/ profiteering—media reports)		31	2.1	0	0.0	0	0.0	0.092 (0.000)
Try it out (curiosity motive)		0	0.0	0	0.0	4	0.4	0.054 (0.026)

No need present		6	0.4	0	0.0	0	0.0	0.040 (0.128)
Rare blood type		0	0.0	1	0.8	14	1.6	0.096 (0.000)
Advertisement/Campaign/Phone call/ Appeal		0	0.0	2	1.5	26	2.9	0.131 (0.000)
• (Missing) previous experience or habit		11	0.7	0	0.0	45	5.0	0.142 (0.000)
Absence of disadvantages		2	0.1	1	0.8	17	1.9	0.094 (0.000)
D07 Benefits/ Incentives	Total n/ domain (D06)	52	3.4	7	5.3	164	18.4	0.249 (0.000)
Blood donor card		0	0.0	0	0.0	3	0.3	0.047 (0.064)
• Compensation		3	0.2	2	1.5	61	6.8	0.196 (0.000)
• Determination of the blood type		0	0.0	0	0.0	14	1.6	0.101 (0.000)
• Health check/screening		0	0.0	0	0.0	21	2.4	0.124 (0.000)
• Exemption from work/school		0	0.0	0	0.0	9	1.0	0.081 (0.000)
• Other services		0	0.0	0	0.0	14	1.6	0.101 (0.000)
D08 Conditions	Total n/ domain (D07)	3	0.2	2	1.5	107	12.0	0.271 (0.000)
• For personal need only		4	0.3	0	0.0	4	0.4	0.020 (0.598)
• Only for family/important others/if person is known		7	0.5	0	0.0	3	0.3	0.018 (0.670)
• In case of a disaster/emergency/ personal experience		5	0.3	0	0.0	9	1.0	0.046 (0.067)
D09 Psychological Aspects	Total n/ domain (D08)	16	1.1	0	0.0	16	1.7	0.041 (0.126)
• Will be made up for/is planned		23	1.5	1	0.8	0	0.0	0.075 (0.001)
• Not ready yet		18	1.2	1	0.8	0	0.0	0.065 (0.005)
• No interest/no wil		126	8.4	1	0.8	0	0.0	0.187 (0.000)
• Indifference/passivity/negligence/ comfort/no desire		53	3.5	0	0.0	1	0.1	0.117 (0.000)
Not thought about it		161	10.7	0	0.0	0	0.0	0.216 (0.000)
• Social aspect/peer group movement/ personal influence and advice		3	0.2	0	0.0	42	4.7	0.163 (0.000)

• Refusal to donate blood		10	0.7	0	0.0	0	0.0	0.052 (0.032)
D10 Missing points of contact	Total n/ domain (D09)	386	25.8	3	2.4	43	4.8	0.278 (0.000)
No request/appellation/call		26	1.7	0	0.0	1	0.1	0.078 (0.000)
Missing reason/occasion		18	1.2	0	0.0	0	0.0	0.070 (0.002)
For no reason		23	1.5	0	0.0	5	0.6	0.050 (0.040)
Misc	Total n/ domain (D10)	67	4.4	0	0.0	6	0.7	0.114 (0.000)
Other		13	0.9	1	0.8	11	1.2	0.018 (0.657)
Do not know		24	1.6	0	0.0	5	0.6	0.053 (0.031)
Not specified		162	10.8	16	12.0	29	3.2	0.134 (0.000)
	Total n/ domain (D11)	199	13.3	17	12.8	45	5.0	0.129 (0.000)

(b)

Domain (D)/Main Category	Subcategory	Non-Intenders (Next 12 Months)		Intenders (Next 12 Months)		Effect Size
D01 Ineligibility		N	%	n	%	Phi
Unspecific reason		20	1.2	0	0.0	−0.059 (0.003)
• Specific reason	Age	284	16.5	5	0.7	−0.225 (0.000)
	Health	337	19.6	3	0.4	−0.254 (0.000)
	Pregnancy	7	0.4	0	0.0	−0.035 (0.083)
• Specific reason	Other	14	0.8	1	0.1	−0.040 (0.049)
D02 Impact/Effect	Total n/ domain (D01)	634	36.8	9	1.2	−0.371 (0.000)
• Physical consequences	Positive	2	0.1	12	1.6	0.092 (0.000)
	Negative—health	62	3.6	2	0.3	−0.096 (0.000)
	Negative—risk	29	1.7	0	0.0	−0.071 (0.000)
• Mental/psychological wellbeing	Positive	0	0.0	19	2.6	0.135 (0.000)
	Negative	6	0.3	0	0.0	−0.032 (0.109)
D03 Fears/Aversion	ToTotal n/ domain (D02)	95	5.5	31	4.2	−0.027 (0.181)
• Fears	Fears in general	40	2.3	0	0.0	−0.084 (0.000)

	Fear of the needle	80	4.6	0	0.0	−0.120 (0.000)
	Fear of blood	5	0.3	0	0.0	−0.030 (0.144)
	Fear to donate	20	1.2	0	0.0	−0.059 (0.003)
	Fear of pain	2	0.1	0	0.0	−0.019 (0.355)
	Other	11	0.6	1	0.1	−0.033 (0.102)
• Aversion	Needles	21	1.2	0	0.0	−0.061 (0.003)
	Blood	27	1.6	0	0.0	−0.069 (0.001)
	Other	21	1.2	2	0.3	−0.045 (0.026)
D04 Obstacles/Barriers	Total n/ domain (D03)	205	11.9	3	0.4	−0.189 (0.000)
• Lack of information		27	1.6	5	0.7	−0.036 (0.076)
• No opportunity/Lack of possibilities		18	1.0	10	1.4	0.014 (0.500)
• Organization/effort		27	1.6	0	0.0	−0.069 (0.001)
• Time/Lack of time		128	7.4	20	2.7	−0.091 (0.000)
• Personal reasons		33	1.9	3	0.4	−0.057 (0.004)
D05 Norms	Total n/ domain (D04)	209	12.1	36	4.9	−0.111 (0.000)
• Reciprocity	General (if no other category)	2	0.1	4	0.5	0.040 (0.049)
	Future-orientated	8	0.5	51	6.9	0.194 (0.000)
	Past-orientated (self)	1	0.1	8	1.1	0.078 (0.000)
	Past-orientated (friends and family)	0	0.0	1	0.1	0.031 (0.126)
• Altruism		24	1.4	297	40.4	0.530 (0.000)
• Obligation/self-evident/social conscientiousness and responsibility		7	0.4	69	9.4	0.238 (0.000)
• Religious reasons		7	0.4	1	0.1	−0.022 (0.281)
• Personal belief		4	0.2	16	2.2	0.099 (0.000)
• Important/necessary/meaningful/good cause		3	0.2	77	10.5	0.266 (0.000)
• Vocational affiliation		0	0.0	8	1.1	0.087 (0.000)
• Surrendering responsibility		25	1.5	0	0.0	−0.066 (0.001)
	Total n/ domain (D05)	77	4.5	474	64.5	0.659 (0.000)

D06Image and Experience		N	%	n	%	Phi
• Compensate for lack of blood products /Support the health care system		2	0.1	68	9.3	0.251 (0.000)
• Lack of trust (rip-off/crime/fraud/ profiteering—media reports)		34	2.0	3	0.4	−0.059 (0.004)
• Try it out (curiosity motive)		0	0.0	2	0.3	0.044 (0.030)
• No need present		8	0.5	1	0.1	−0.025 (0.217)
• Rare blood type		2	0.1	11	1.5	0.087 (0.000)
• Advertisement/ Campaign/Phone call/ Appeal		3	0.2	4	0.5	0.032 (0.115)
• (Missing) previous experience or habit		16	0.9	38	5.2	0.132 (0.000)
• Absence of disadvantages		3	0.2	30	4.1	0.155 (0.000)
D07 Benefits/ Incentives	Total n/ domain (D06)	68	4.0	149	20.3	0.263 (0.000)
• Blood donor card		0	0.0	0	0.0	-
• Compensation		19	1.1	19	2.6	0.055 (0.006)
• Determination of the blood type		0	0.0	1	0.1	0.031 (0.126)
• Health check/ screening		0	0.0	11	1.5	0.103 (0.000)
• Exemption from work/school		0	0.0	0	0.0	–
• Other services		3	0.2	5	0.7	0.041 (0.044)
D08 Conditions	Total n/ domain (D07)	22	1.3	33	4.5	0.099 (0.000)
• For personal need only		5	0.3	2	0.3	−0.002 (0.938)
• Only for family/important others/if person is known		11	0.6	5	0.7	0.002 (0.907)
• In case of a disaster/emergency/ personal experience		8	0.5	4	0.5	0.005 (0.795)
D09 Psychological Aspects	Total n/ domain (D08)	22	1.3	11	1.5	0.009 (0.666)
• Will be made up for/ is planned		7	0.4	29	3.9	0.135 (0.000)
• Not ready yet		17	1.0	2	0.3	−0.037 (0.064)
• No interest/no will		117	6.8	4	0.5	−0.132 (0.000)

• Indifference/passivity/negligence/ comfort/no desire		36	2.1	3	0.4	−0.062 (0.002)
• Not thought about it		42	2.4	5	0.7	−0.059 (0.004)
• Social aspect/peer group movement/ personal influence and advice		5	0.3	18	2.4	0.103 (0.000)
• Refusal to donate blood		8	0.5	0	0.0	−0.037 (0.064)
D10 Missing points of contact	Total n/ domain (D09)	229	13.3	59	8.0	−0.075 (0.000)
• No request/ appellation/call		11	0.6	4	0.5	−0.006 (0.783)
• Missing reason/ occasion		15	0.9	1	0.1	−0.042 (0.038)
• For no reason		19	1.1	3	0.4	−0.034 (0.094)
Misc	Total n/ domain (D10)	44	2.6	8	1.0	−0.047 (0.021)
• Other		16	0.9	9	1.2	0.013 (0.504)
• Do not know		26	1.5	3	0.4	−0.047 (0.021)
• Not specified		241	14.0	48	6.5	−0.106 (0.000)
	Total n/ domain (D11)	283	16.4	60	8.1	−0.109 (0.000)

Key: "(a): * Categories are not mutually exclusive. ** In bold print: category/domain percentages >5%. *** Total number per domain indicates the number of cases with at least one category in this domain. **** Effect size interpretation: Cramer's V > 0.10 = small effect; Phi > 0.30 = medium effect; Phi > 0.50 = large effect (Cohen, 1988); (b): * Categories are not mutually exclusive. ** In bold print: category/domain percentages >5%. *** Total number per domain indicates the number of cases with at least one category in this domain. **** Effect size interpretation: Phi > 0.10 = small effect; Phi > 0.30 = medium effect; Phi > 0.50 = large effect (Cohen, 1988)."

Source: Greffin K, Schmidt S, Schönborn L, Muehlan H. "Blood for Blood"? Personal Motives and Deterrents for Blood Donation in the German Population. *International Journal of Environmental Research and Public Health*. 2021; 18(8):4238. https://doi.org/10.3390/ijerph18084238:

The understanding of the factors promoting or deterring blood donation is complex. Given that the population is dynamic with changing attitudes, access to knowledge, and associated fears, a global strategy that amalgamates all these aspects is warranted. Mostly, the aspect of "fear" needs to be addressed.

"Building a sustainable base of safe blood donors requires a long-term approach, which involves not only the establishment of an effective voluntary blood donor program but also an improved public awareness of the importance of blood donation as a social norm."

— The WHO and the International Federation of Red Cross and Red Crescent Societies (IFRC): Global status report on blood safety and availability 2016. Geneva: World Health Organization; 2017.

In a qualitative analysis by Mugion et al. (2021), other important aspects that were highlighted included the waiting time for donation, the cleanliness of transfusion centers, and the attitude of the medical staff. Two examples of feedback opinions collected by the researchers have been presented below (Guglielmetti Mugion, et al., 2021):

"The transfusion centers must guarantee easy access to the donation center and easy-to-find information about places and times. Transfusion centers must also be safe and therefore guarantee hygiene and qualified staff."

When I donate, I pay attention to whether the medical staff are friendly and qualified, provide polite treatment, and to tangible aspects such as the cleanliness of the facilities."

Other significant opinions from non-donors that emerged were the "sight of blood" and "the lack of awareness" – a few examples are cited below as collected by the researchers (Guglielmetti Mugion et al., 2021):

"The sight of blood is unpleasant, and I'm afraid of needles and of infectious disease transmission."

"Given the lack of blood, the advertising campaign should be increased, especially in the summer, given the low number of donors."

"In small towns, people are not informed about the importance of donation. Donation initiatives are not advertised. Information is often not provided on the places, days, and times to donate."

An interesting set of motivating factors was compiled by the authors and has been presented in Table 4.4 (Guglielmetti Mugionet al., 2021).

Table 4.4. A few motivating factors for blood donation

Donors (a)					
Motivation for First Donation	Frequency	%	Favorite Location for Donating Blood	Frequency	%
Individual choice	64	37%	School/universities	77	45%

Educational initiatives	29	17%	City center areas (plaza or parks)	42	24%
Accompanying relatives/friends	23	13%	In the office	24	14%
Sensitive campaign	23	13%	Sport center	12	7%
Blood need for relatives/friends	14	8%	Hospitals	7	4%
Being a parent of a donor	8	5%	Blood donation center	2	1%
Altruism	6	3%	Churches	2	1%
Being a friend of a donor	2	1%	Everywhere	2	1%
Free breakfast	1	1%	Place with parking	2	1%
Familiar education	1	1%	Associations	1	1%
Personal motivation	1	1%	Mall	1	1%
Personal satisfaction	1	1%	Game rooms	1	1%
Non-Donors (b)					
Motivations for starting the donation	Frequency	%	Favorite location to start donating	Frequency	%
My parents/friend need blood	47	54%	Doesn't matter	45	52%
Sensitive campaigns	18	21%	Blood donation center	34	39%
Supporting those who are in need	9	10%	Association center	5	6%
For ethical reasons	6	7%	Mobile blood station	3	3%
Educational event at school/ university	3	3%			
Accompanying parents/friends	3	3%			
Overcoming my fears	1	1%			

Source: Guglielmetti Mugion, R., Pasca, M.G., Di Di Pietro, L. *et al.* Promoting the propensity for blood donation through the understanding of its determinants. *BMC Health Serv Res* **21**, 127 (2021). https://doi.org/10.1186/s12913-021-06134-8:

Cognitive maps made by the researchers to comprehend the ideas of donors and non-donors have been presented in Figures 4.1 and 4.2 to increase our understanding.

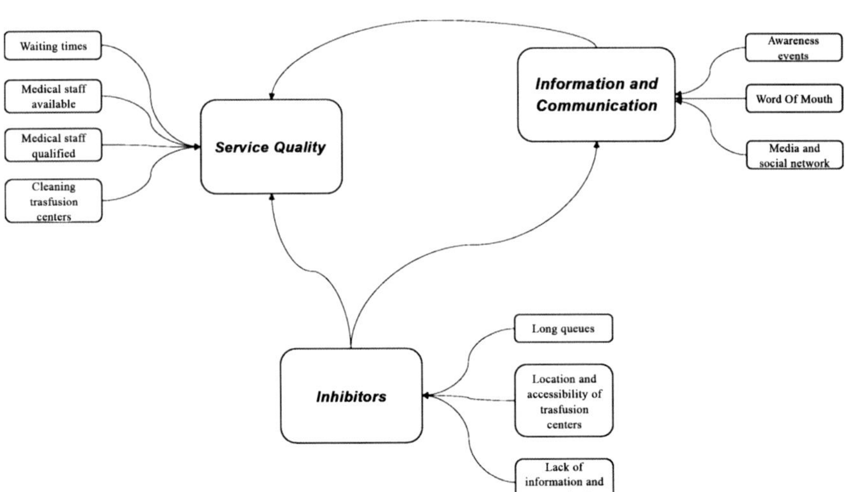

Figure 4.1. Cognitive map of factors associated with blood donors.

Source: Guglielmetti Mugion, R., Pasca, M.G., Di Di Pietro, L. *et al.* Promoting the propensity for blood donation through the understanding of its determinants. *BMC Health Serv Res* **21**, 127 (2021). https://doi.org/10.1186/s12913-021-06134-8:

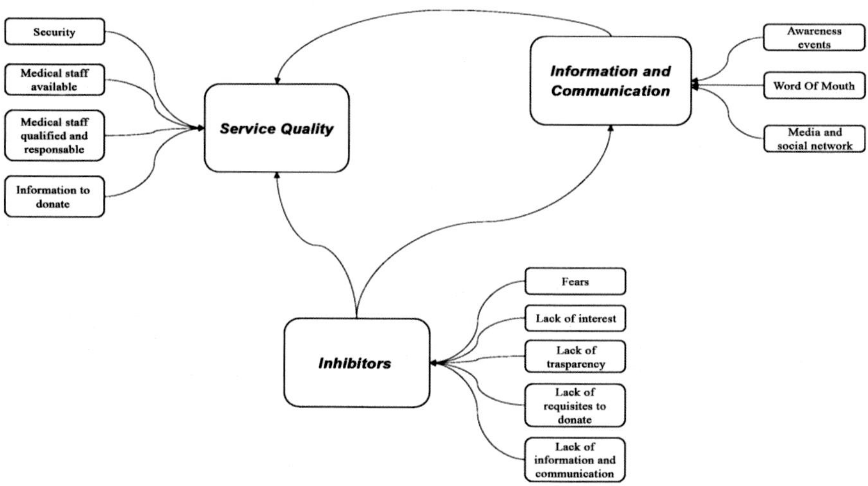

Figure 4.2. Cognitive map of factors associated with non-blood donors.

Source: Guglielmetti Mugion, R., Pasca, M.G., Di Di Pietro, L. *et al.* Promoting the propensity for blood donation through the understanding of its determinants. *BMC Health Serv Res* **21**, 127 (2021). https://doi.org/10.1186/s12913-021-06134-8:

Further, studies in different cohorts reveal that the strategies to recruit donors should be dynamic as a slew of factors are involved ranging from education to socioeconomic status to one's current mindset.

"A bottle of blood saved my life. Was it yours?"

—https://tsbtc.tripura.gov.in/Blood-Donors-Slogans

4.2. ARTIFICIAL INTELLIGENCE TO PREDICT BLOOD DONATION?

"The goal of artificial intelligence is to build machines that can think and learn like humans, but the ultimate objective is to build machines that are even better than humans at thinking and learning."

—Sebastian Thrun

The relevance of a computer algorithm to match the judgment of a human with respect to blood transfusion was probed by researchers Yao and team (2019). This becomes pertinent given that there are reports of "over-transfusions" being prescribed in clinical practice (Goodnough and Murphy, 2014). According to Rajendran and Ravindran (2019), inventory management is warranted for blood management, especially for short-lived cells like platelets.

To see how complex blood management is, let us consider a retrospective, multicenter, cross-sectional case review study on nine hospitals: 3 tertiary hospitals in each of the three regions in Zhejiang, i.e., south, north, and central regions on adult patients who received blood transfusions during surgical care in the years 2007 and 2011. The analysis revealed that "a smaller portion of patients received allogeneic blood transfusions in 2011 than in 2007 (49.9% vs. 74.8%, $p<0.001$), with a relatively smaller amount of mean allogeneic RBC use in 2011 than in 2007 (671 ± 472 ml vs. 702 ± 503 ml, $p<0.001$) "during surgery. Further, the "proportion of intraoperative blood transfusions without appropriate assessment in 2011 was significantly lower than in 2007 (intraoperatively: 32.5% vs. 63.0% [$p<0.001$]; postoperatively: 28.7% vs. 40.0% [$p<0.001$])." In order to further illuminate that blood management is complex, the table below highlights the findings of the study: (Yao, Li, Wang, et al., (2018) Improvements in blood transfusion management: cross-sectional data analysis from nine hospitals in Zhejiang, China. *BMC Health Serv Res.*, *18*, 856. https://doi.org/10.1186/s12913–018–3673-x: "*Open Access: This article is distributed*

Table 4.5. A case study from China documenting significant improvements in the quality of blood transfusion management

	2007	2011	*P*-Value
Patient receiving surgery (N)	111,655	158,766	
Patient receiving blood transfusion during hospital care (N; %)	6,093; 5.4%	9,646; 6.1%	<0.001
Patient receiving perioperative allogeneic RBC transfusion (N; %)	6,000; 5.4%	7,468; 4.7%	< 0.001
Intra-operative			
Intra-operative blood loss (ml, mean ± SD)	873 ± 1,000	722 ± 812	< 0.001
Intra-operative blood transfusion (N; %)	4,560; 74.8%	4,812; 49.9%	< 0.001
Intra-operative blood transfusion (Volume in ml, mean ± SD)	702 ± 503	671 ± 472	< 0.001
Pre-transfusion hemoglobin test (N; %)	1,688; 37.0%	3,250; 67.5%	< 0.001
Transfusion hemoglobin (g/dl, mean ± SD)	93.1 ± 26.7	89.1 ± 24.3	0.002
Pre-transfusion Hct (%, mean ± SD)	28.5 ± 8.1	26.9 ± 7.0	< 0.001
Appropriateness (N; %)	1,063; 63%	2,535; 78%	< 0.001
Post-operative			
Post-operative blood loss (ml, mean ± SD)	500 ± 609	491 ± 621	0.46
Post-operative blood transfusion (N; %)	3,164; 51.9%	4,268; 44.2%	< 0.001
Post-operative blood transfusion (Volume in ml, mean ± SD)	712 ± 562	632 ± 517	< 0.001
Pre-transfusion hemoglobin test (N; %)	1,898; 60.0%	3,042; 71.3%	< 0.001
Transfusion hemoglobin (g/dl, mean ± SD)	89.9 ± 22.5	85.4 ± 20.2	< 0.001

Pre-transfusion Hct (%, mean ± SD)	26.8 ± 6.7	25.5 ± 6.0	< 0.001
Appropriateness (N; %)	1,340; 70.6%	2,386; 78.4%	< 0.001
Cell salvage or ANH			
Acute normovolemic hemodilution (N; %)	3; 0.1%	894; 9.3%	< 0.001
Intra-operative cell salvage (N; %)	383; 6.3%	2,665; 27.6%	< 0.001
Average volume of cell salvage per patient intraoperative (ml)	63	172	
Post-operative cell salvage (N; %)	17; 0.3%	41; 0.4%	< 0.001

With this background, Yao et al. (2019) checked the appropriateness of blood transfusion for the condition of a large volume of clinical cases using Restricted Boltzmann Machines (RBM) for the artificial neural network (ANN). Using a dataset with 15,000 surgical cases collected from 9 hospitals in Zhejiang, China in 2007 and 2011, the researchers employed a multilayer perceptron neural network (MLPNN) to screen the appropriateness of blood transfusions. While expert anesthesiologists rated 3604 cases as appropriate and 1342 as inappropriate from 4946 patients who received intra-operative blood transfusions, on the basis of WHO guidelines, the result for the AI tool was a 96.8% accuracy rate in matching human judgment: 99% of the computer's decisions matched the experts on the appropriate cases and 90.9% matched on the inappropriate cases.

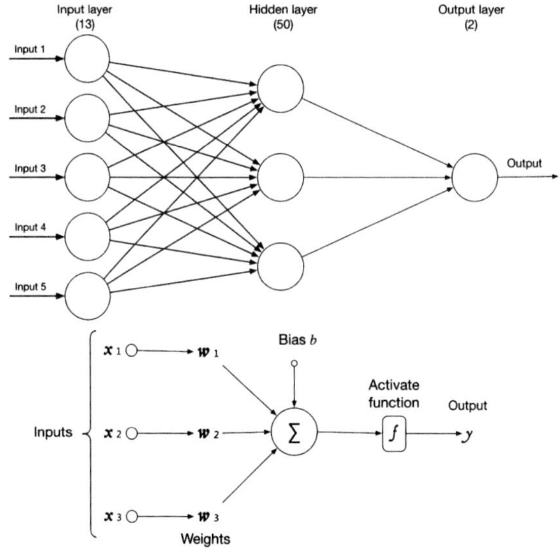

Table 4.6. A study on how an algorithm can match transfusion

The developed algorithm and the outcome are given below as depicted by the authors (Yao et al., 2019: *Open Access: This article is distributed under the terms of the Creative Commons Attribution 4.0 International License (http://creativecommons.org/licenses/by/4.0/), which permits unrestricted use, distribution, and reproduction in any medium, provided you give appropriate credit to the original author(s) and the source, provide a link to the Creative Commons license, and indicate if changes were made. The Creative Commons Public Domain Dedication waiver (http://creativecommons.org/publicdomain/zero/1.0/) applies to the data made available in this article, unless otherwise stated*):

Physician Judgment		General Classification	
Need Transfusion	**No Need Transfusion**		
Computer Report			
Yes	True positive 3,569/3,604 (99.0%)	False negative 122/1,342 (9.1%)	
No	False positive 35/3,604 (1.0%)	True negative 1,220/1,342 (90.9%)	
			96.8% (3,569 + 1,220)/4,946 × 100%

Source: Yao, Y., Cifuentes, J., Zheng, B. *et al.* Computer algorithm can match physicians' decisions about blood transfusions. *J Transl Med* **17**, 340 (2019). https://doi.org/10.1186/s12967-019-2085-y

In an interesting article published in Symmetry, the authors usedAI to predict the intention to re-donate among blood donors. The theory of planned behavior (TPB), which is commonly employed to model the blood donation intent, was used by the authors with an in-person questionnaire for users (adults) in two health centers in Valdivia (Chile). The idea of the TPB is illustrated in the Figure 4.3 as depicted by the researchers (Salazar-Concha C, Ramírez-Correa P. Predicting the Intention to Donate Blood among Blood Donors Using a Decision Tree Algorithm. *Symmetry*. 2021; 13(8):1460: © 2021 by the authors. Licensee MDPI, Basel, Switzerland. This article is an open access article distributed under the terms and conditions of the Creative Commons Attribution (CC BY) license: https://creativecommons.org/licenses/by/4.0/).

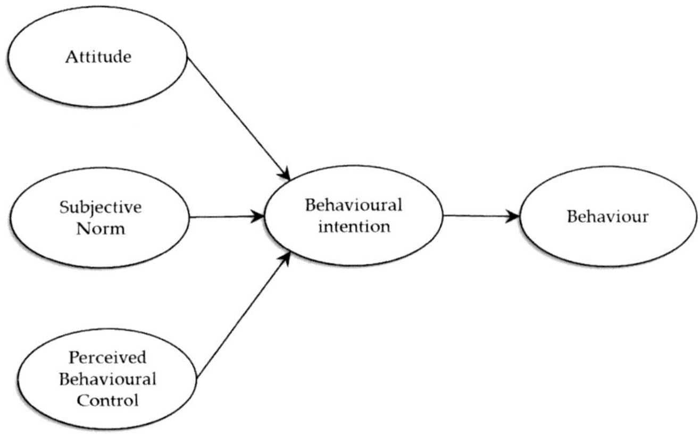

Figure 4.3. *The idea of the TPB.*

Subsequently, using a decision tree, the researchers then predicted the intended blood donation. They reported an accuracy of 84.17% in predicting the intention to donate blood again. The outline and the outcome are depicted in the Figures 4.4 as reported by the authors (Salazar-Concha C, Ramírez-Correa P. Predicting the Intention to Donate Blood among Blood Donors Using a Decision Tree Algorithm. *Symmetry*. 2021; 13(8):1460: © 2021 by the authors. Licensee MDPI, Basel, Switzerland. This article is an open access article distributed under the terms and conditions of the Creative Commons Attribution (CC BY) license: https://creativecommons.org/licenses/by/4.0/):

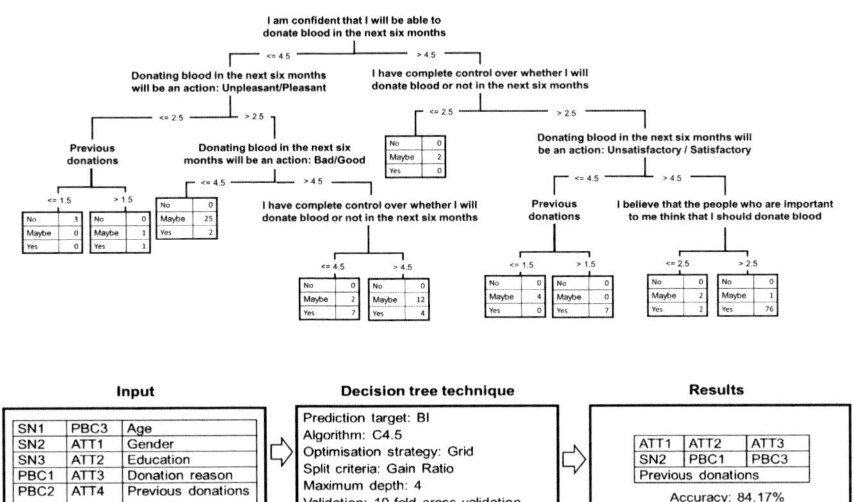

Figure 4.4. *The outline and the outcome.*

Key: "Attitude (ATT, positive or negative assessments of a behavior), subjective norm (SN, general perception of social pressure to adopt or not adopt a behavior), and perceived behavioral control (PBC, perceived control over or a capacity to perform a behavior)."

Given the "tight balance" of blood during exigencies such as natural calamities and human-caused disasters, Wu et al (2022) reported the use of machine learning to predict and rank the willingness to donate blood and meet precision blood recruitment. Interestingly, short message service (SMS) recruitment was employed to collect data from experienced blood donors (95,476 recruitments) at the Yangzhou Blood Station from February 1, 2019, to September 30, 2021. Machine learning-based recruitment models were then developed with the PyCharm-Python Environment and assessed with relevant parameters like Area under the receiver operating characteristic curve (AUC), accuracy, precision, etc. The approach employed as depicted by the researchers is illustrated in Figure 4.5 (Wu, Hy., Li, Zg., Sun, Xk. et al., 2022: *Sci Rep 12, 19165: Open Access: This article is licensed under a Creative Commons Attribution 4.0 International License, which permits use, sharing, adaptation, distribution and reproduction in any medium or format, as long as you give appropriate credit to the original author(s) and the source, provide a link to the Creative Commons license, and indicate if changes were made. The images or other third party material in this article are included in the article's Creative Commons license, unless indicated otherwise in a credit line to the material. If material is not included in the article's Creative Commons license and your intended use is not permitted by statutory regulation or exceeds the permitted use, you will need to obtain permission directly from the copyright holder. To view a copy of this license, visit http://creativecommons.org/licenses/by/4.0/*). Predicting willingness to donate blood based on machine learning: two blood donor recruitments during COVID-19 outbreaks).

Of the seven models developed, the best performance was reported for the eXtreme Gradient Boosting (XGBoost) and Support Vector Machine (SVM) models: mean (95% CI) with the highest AUC: 0.809 (0.806–0.811), accuracy: 0.815 (0.812–0.818), precision: 0.840 (0.835–0.845), and F1 score of XGBoost: 0.843 (0.840–0.845) and recall of SVM: 0.991 (0.988–0.994). Furthermore, the hit rate of the XGBoost model alone was 1.25 times higher than that of the conventional method, while that of the combined XGBoost and SVM models was found to be 1.80 times higher than that of the conventional method as a control in two recruitments. Additionally, the hit rate of the "high willingness to donate group" was 1.96 times higher than that of the "low willingness to donate group." The test accuracies of the developed AI systems as reported by the researchers are given in Figure 4.6 (Wu, Hy., Li, Zg., Sun, Xk. et al.

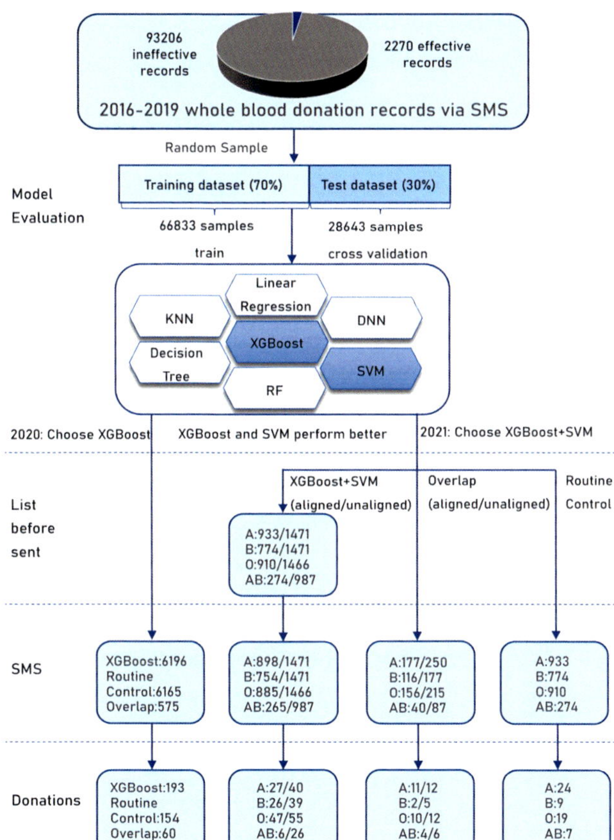

Figure 4.5. An example of the use of machine learning to predict and rank the willingness to donate blood and meet precision blood recruitment

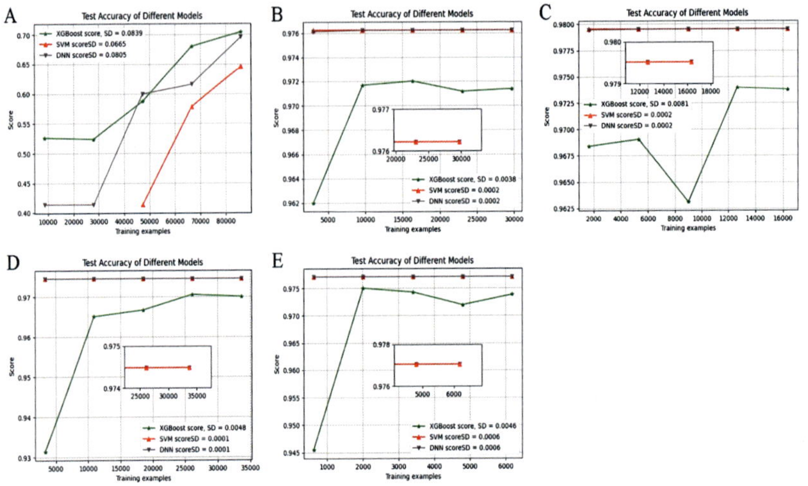

Figure 4.6. Test accuracies of the developed AI systems by Wu et al (2022)

Key: The accuracy calculated by XGBoost, SVM, and DNN algorithms, respectively: (A) in the full ABO dataset, (B) in the A blood group dataset, (C) in the B blood group dataset, (D) in the O blood group dataset, and (E) in the AB blood group dataset.

Interestingly, both the conventional method and the machine learning (ML) approach did not recommend blood donors with intervals longer than 1.5 years. However, the proportion of the ML model-recommended donors with intervals within 1 year in 2020 was 93.3%, compared to 77.9% for the conventional method. The details of the blood donor recruitment during the COVID-19 pandemics in 2020 and 2021 as classified by the ML approach are presented below as put forth by the researchers (Wu, Hy., Li, Zg., Sun, Xk. et al., 2022: *Sci Rep 12, 19165: Open Access: This article is licensed under a Creative Commons Attribution 4.0 International License, which permits use, sharing, adaptation, distribution and reproduction in any medium or format, as long as you give appropriate credit to the original author(s) and the source, provide a link to the Creative Commons license, and indicate if changes were made. The images or other third party material in this article are included in the article's Creative Commons license, unless indicated otherwise in a credit line to the material. If material is not included in the article's Creative Commons license and your intended use is not permitted by statutory regulation or exceeds the permitted use, you will need to obtain permission directly from the copyright holder. To view a copy of this license, visit http://creativecommons.org/licenses/by/4.0/*).

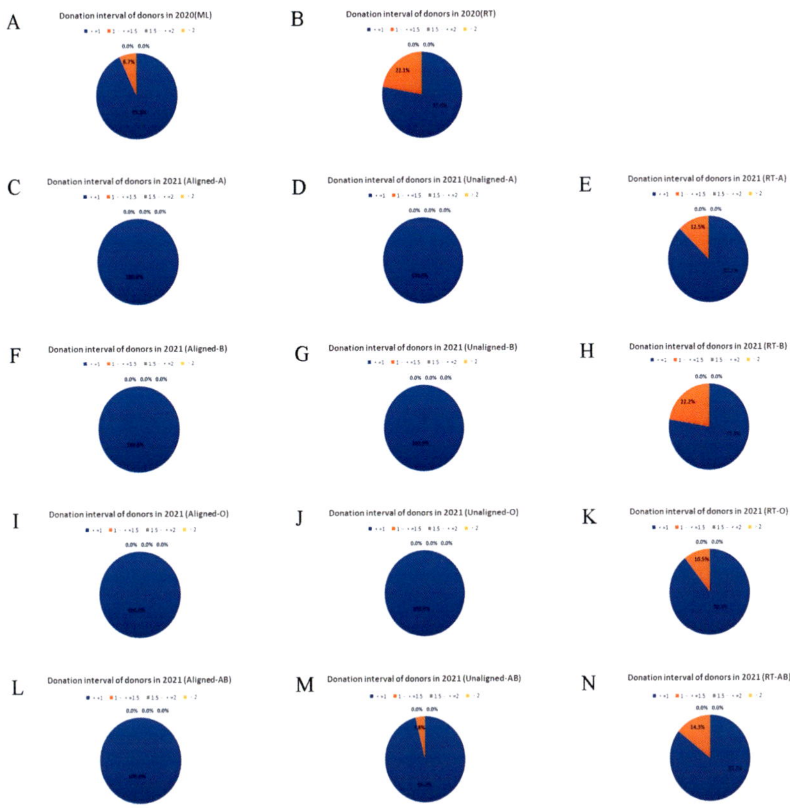

Figure 4.7. ML for blood donor recruitment during the COVID-19 pandemics in 2020 and 2021 by Wu et al (2022)

Key: Donation interval of donors (A) in 2020 (ML). (B) in 2020 (RT). (C) in 2021 (Aligned-A). (D) in 2021 (Unaligned-A). (E) in 2021 (RT-A). (F) in 2021 (Aligned-B). (G) in 2021 (Unaligned-B). (H) in 2021 (RT-B). (I) in 2021 (Aligned-O). (J) in 2021 (Unaligned-O). (K) in 2021 (RT-O). (L) in 2021 (Aligned-AB). (M) in 2021 (Unaligned-AB). (N) in 2021 (RT-AB).

The case studies presented above outline the promise of using AI/ML tools in the field of blood management. However, intensified research and more studies are warranted for the suitability of their application in actual clinical translation.

"It's going to be interesting to see how society deals with artificial intelligence, but it will definitely be cool."

—Colin Angle

4.3. MANUFACTURING BLOOD?

In a landmark study published by Giarratana et al. (2013), a large-scale *ex vivo* production approach for mature human red blood cells (RBCs) from hematopoietic stem cells was reported. The researchers amplified CD34+ stem cells (up to 1.95 × 10^6-fold) by using cytokines and co-culture on stromal cells. These cells underwent terminal differentiation into fully mature, functional RBCs (100%) and survived in nonobese diabetic/severe combined immunodeficient mice like native RBC cells. This report, among others, potentially paves the way for manufacturing blood to address the shortage in blood transfusion.

In addition to the understanding of the process of *in vivo* RBC generation in the blood (the process is called erythropoiesis), researchers are also attempting to mimic what happens under *in vitro* conditions. Cells including CD34+ hematopoietic stem/progenitor cells (HSPCs), stem cells (embryonic and induced pluripotent cells), and erythroid precursors that have been immortalized have been assayed along with growth factors for this purpose (Singh, Saini, Tsuji, Sharma, & Chandra, 2014). These approaches for expanding RBCs have been summarized in the Figure 4.8 (Copyright © 2014 Singh, Saini, Tsuji, Sharma, and Chandra).

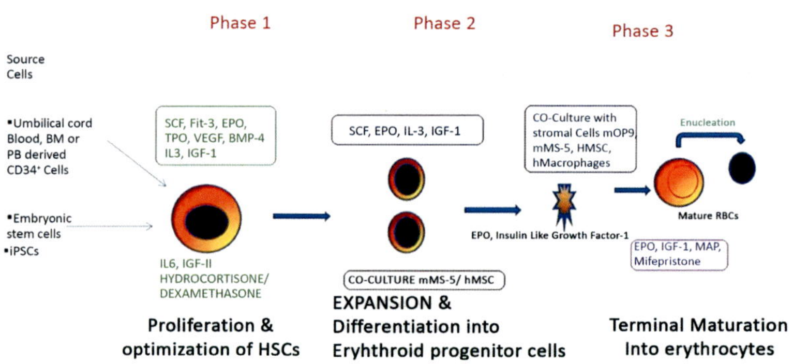

Figure 4.8. *Approaches for expanding RBCs.*

Table 4.7 summarizes various research approaches employed for *in vitro* RBC production.

Table 4.7. Various Research Approaches Employed for *in vitro* RBC

1	CB CD34+ 21 days	FLT-3, SCF, TPO, EPO, IGF-1	NO	NA	NA	4% *in vitro* 99% *in vivo*	No	Basis for transfusion potential of *ex vivo* cultured RBCs in animal model	Neildez-Nguyen et al., 2002
2	Cord blood 21 days	SCF, IL-3, EPO hydrocortisone	Murine stromal mMS-5	1.95×10^6	4.6/CB	95%	No	First *in vitro* production of RBCs	Giarratana et al., 2005
3	Cord blood 60 days	EPO, SCF, IGF-1 dex and lipid mix	No	10^9	-NA	~100%	Yes	Prolonged expansion protocol up to 60 days for adult globin switching	Leberbauer et al., 2005
4	Cord blood 20 days	SCF, IL-3, EPO VEGF, IGF-2	No	7.2×10^5	10^4/CB	77.5%	Yes	Low expansion and enucleation rates	Miharada et al., 2006
5	Cord blood 21 days	SCF, IL-3, EPO, TPO, Flt-3	Human MSCs	8×10^3	0.02/CB	64%	No	Less allogenicity, lower expansion, and enucleation rates, replaced BM derived feeder cells with UCB derived cells	Baek et al., 2008
6	Cord blood 38 days	SCF+Flt-3+TPO IL-3+EPO	hTERT+ Macrophages	3.5×10^6	8.8/CB	100%	Yes	Impossible to scale up	Fujimi et al., 2008
7	hESCs 59 days	Hydrocortisone, IL3, BMP-4, SCF, EPO, IGF-1	hMSCs, mMS-5 FH-B-hTERT	4×10^7 cells	NA	(6.5%, starting from CD34+ cells)	Yes	Globin switching	Qiu et al., 2008
8	hESCs 42 days	SCF, Epo, BMP-4, VEGF, bFGF, TPO, FLT3 L	MEFs, OP9	10^{10}–10^{11} cells/6-well plate of hESCs	NA	10–65%	No	Functional oxygen carrying capacity of ESCs derived RBCs	Lu et al., 2008
10	hiPSCs,(MR90, FD-136) hESCs(H1l) 46 days	SCF,TPO, FLT3-L, TPO, BMP-4; VEGF- IL-3, IL-6		hIPSCs 4.4×10^8 hESCs 35×10^8	NA	4–10% 52–66%	No	First time complete differentiation of hiPSCs cells into definitive erythrocytes capable of maturation up to enucleated RBCs (fetal hemoglobin in a functional tetrameric form)	Lapillonne et al., 2010
11	Cord blood 33 days	SCF, IL-3, EPO hydrocortisone	No	2.25×10^8	~500 units/UCB	>90%	No	First serum-free culture, first demonstration of RBC culture in a large-scale bioreactor	Timmins et al., 2011
12	Cord blood 18 days	SCF, EPO hydrocortisone	No	4.3×10^7	75/CB	70%	Yes	Best yield to date	Giarratana et al., 2011
13	hiPSCs 59 days	Hydrocortisone, IL-3 BMP-4, Flt3L, SCF, EPO, IGF-1	hMSCs/Matrigel. hTERT immortalized fetal liver cells	0.5–8×10^6	NA	NA	Yes	Production of large number of erythroid cells with embryonic and fetal-like characteristics regardless of the age of donor tissue	Chang et al., 2011
14	hESCs hiPSCs 60–125 days	Iron saturated transferrin, dexamethasone, insulin, SCF, EPO, TPO, IL-3, IL-6	MEFs, mOP9 mMS-5	2×10^5 ESCs	NA	2–10%	Yes	RBCs production from transgenic and transgene-free iPSCs using the OP9 coculture method with efficiency comparable to hESCs	Dias et al., 2011
15	hiPSCs 52 days	holo-human transferrin recombinant human insulin heparin, and 5% human plasma SCF, TPO, FLT3 ligand, BMP4, VEGF- n-3, IL-Epo	MEFs	15–28.3×10^8	NA	20–26% RBC and 74–80% orthochromatic erythroblasts	5–10% Hu plasma	First time in a normal and a pathological erythropoietic differentiation models that hiPSC are intrinsically able to mature into adult hemoglobin synthesizing cells	Kobari et al., 2012

Research is unraveling the potential of "commercially producing RBCs" in the future. While the potential looks promising, certain aspects emerge including strict quality control, looking at how to lower the use of coculture or feeder cells (to avoid potential immunogenic reactions), and further development of more sophisticated tools.

"Life is too complicated to permit a complete understanding through the study of whole organisms. Only by simplifying a biological problem – breaking it down into a multitude of individual problems -- can you get the answers. In 1935, for example, Charles Coryell and I made our discovery about how oxygen molecules are attached to the iron atoms of hemoglobin, not by getting a cow and putting it into our magnetic apparatus, but by getting some blood from the cow and studying this blood."

—Linus Pauling. Interview with Neil A. Campbell, Bioscience, v. 36, no. 11. December 1986.

REFERENCES

1. Bednall, T. C., Bove, L. L., Cheetham, A., & Murray, A. L. (2013). A systematic review and meta-analysis of antecedents of blood donation behavior and intentions. *Social Science & Medicine, 96*, 86–94. https://doi.org/10.1016/j.socscimed.2013.07.022.
2. Giarratana, M. C., Kobari, L., Lapillonne, H., et al. (2005). Ex vivo generation of fully mature human red blood cells from hematopoietic stem cells. *Nature Biotechnology, 23*, 69–74. https://doi.org/10.1038/nbt1047.
3. Goodnough, L. T., & Murphy, M. F. (2014). Do liberal blood transfusions cause more harm than good? *BMJ, 349*, Article g6897.
4. Greffin, K., Schmidt, S., Schönborn, L., &Muehlan, H. (2021). "Blood for Blood"? Personal motives and deterrents for blood donation in the German population. *International Journal of Environmental Research and Public Health, 18*(8), Article 4238. https://doi.org/10.3390/ijerph18084238.
5. GuglielmettiMugion, R., Pasca, M. G., Di Di Pietro, L., et al. (2021). Promoting the propensity for blood donation through the understanding of its determinants. *BMC Health Services Research, 21*, Article 127. https://doi.org/10.1186/s12913-021-06134-8.
6. Li, Z., Lei, S., Li, X., Zhao, Y., Dai, Y., Jin, S., Fu, Q., Cai, X., Lin, Z., & Tu, X. (2021). Blood donation fear, perceived rewards, self-efficacy, and intention to return among whole blood donors in China: A social cognitive perspective. *Frontiers in Psychology, 12*, Article 683709. https://doi.org/10.3389/fpsyg.2021.683709.
7. Rajendran, S., & Ravi Ravindran, A. (2019). Inventory management of platelets along blood supply chain to minimize wastage and shortage. *Computers & Industrial Engineering, 130*, 714–730.
8. Renaux, O., Bouazzi, L., Sanchez, A., Hottois, J., Martin, M.-C., Chrusciel, J., & Sanchez, S. (2022). Impact of promoting blood donation in general practice: Prospective study among blood donors in France. *Frontiers in Public Health, 10*, Article 1080096. https://doi.org/10.3389/fpubh.2022.1080096.
9. Roberts, N., James, S., Delaney, M., & Fitzmaurice, C. (2019). The global need and availability of blood products: A modelling study. *The Lancet. Haematology, 6*(12), e606–e615. https://doi.org/10.1016/S2352-3026(19)30200-5.
10. Salazar-Concha, C., & Ramírez-Correa, P. (2021). Predicting the intention to donate blood among blood donors using a decision tree algorithm. *Symmetry, 13*(8), Article 1460. https://doi.org/10.3390/sym13081460.

11. Singh, V. K., Saini, A., Tsuji, K., Sharma, P. B., & Chandra, R. (2014). Manufacturing blood ex vivo: A futuristic approach to deal with the supply and safety concerns. *Frontiers in Cell and Developmental Biology, 2*, Article 26. https://doi.org/10.3389/fcell.2014.00026.

12. Wu, H. Y., Li, Z. G., Sun, X. K., et al. (2022). Predicting willingness to donate blood based on machine learning: Two blood donor recruitments during COVID-19 outbreaks. *Scientific Reports, 12*, Article 19165. https://doi.org/10.1038/s41598-022-21215-2.

13. Yao, Y. Y., Li, J., Wang, M. C., Chen, Z. H., Wang, W. X., Lei, L. P., Huang, C. S., Yao, M., Yuan, G. H., & Yan, M. (2018). Improvements in blood transfusion management: Cross-sectional data analysis from nine hospitals in Zhejiang, China. *BMC Health Services Research, 18*, Article 856.

14. Yao, Y., Cifuentes, J., Zheng, B., et al. (2019). Computer algorithm can match physicians' decisions about blood transfusions. *Journal of Translational Medicine, 17*, Article 340. https://doi.org/10.1186/s12967-019-2085-y.

Chapter 5

BLOOD MANAGEMENT: RESEARCH INSIGHTS

"*Blood is the very source of existence for all humans. Globally, blood transfusions save millions of lives annually and enhance the life expectancy and quality of life of patients with life-threatening conditions, some of whom require regular transfusions for the rest of their lives. Blood is the most precious gift that anyone can give to another person – the gift of life. A decision to donate blood can save a life, or even several if blood is separated into its components – red cells, platelets, and plasma – which can be used individually for patients with specific conditions. Blood and blood products are required for the treatment of blood dyscrasias, surgeries, obstetric complications, and as a treatment supplement for various diseases.*

Availability of quality-assured, safe blood and blood products in a sustainable way is therefore essential for a resilient health system."

—Dr Poonam Khetrapal Singh,
WHO Regional Director for South-East Asia

5.1. BLOOD SUPPLY MANAGEMENT: A REVIEW OF THE RESEARCH STATUS

The handling of blood delivery and supply chain services is an important aspect of the healthcare system across nations. Stocking an adequate blood supply becomes imperative for efficient healthcare services, especially taking into account environmental, social, and economic aspects. Blood supply chain management (BSCM) entails several aspects including blood donation, transportation, and storage. This becomes complex given that different blood products have varying lifespans and purity must be maintained (Alghamdi, 2023). An operational process for blood supply has been depicted Figure 5.1.

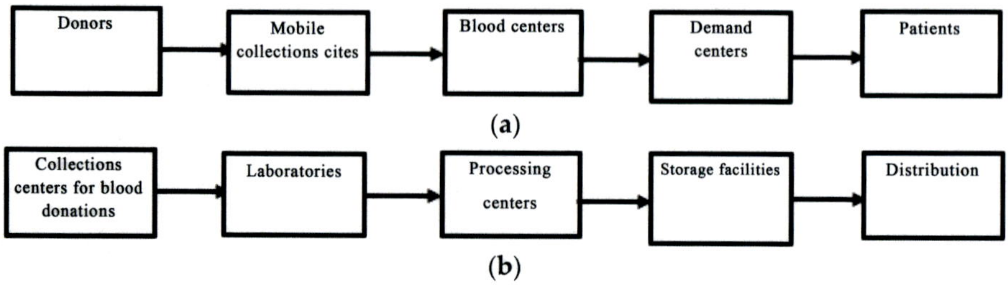

Figure 5.1. *An operational process for blood supply.*

Source: Alghamdi SY. A Review of Blood Delivery for Sustainable Supply Chain Management (BSCM). *Sustainability*. 2023; 15(3):2757. https://doi.org/10.3390/su15032757: © 2023 by the author. Licensee MDPI, Basel, Switzerland. This article is an open access article distributed under the terms and conditions of the Creative Commons Attribution (CC BY) license (**https://creativecommons.org/licenses/by/4.0/**).

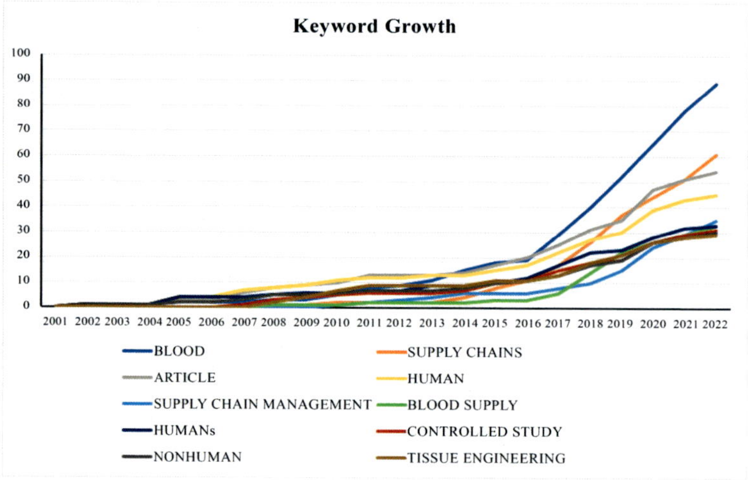

Figure 5.2. Frequency of search words for blood management

Source: Alghamdi SY. A Review of Blood Delivery for Sustainable Supply Chain Management (BSCM). *Sustainability*. 2023; 15(3):2757. https://doi.org/10.3390/su15032757: © 2023 by the author. Licensee MDPI, Basel, Switzerland. This article is an open access article distributed under the terms and conditions of the Creative Commons Attribution (CC BY) license (**https://creativecommons.org/licenses/by/4.0/**).

In order to understand the relevance of blood supply management in research, a review by Alghamdi (2023) has been presented here to illuminate the research trends and scenario. Frequency of keywords "Supply Chain Management," "Supply chain," "Blood Supply," and "Blood" used in association that were obtained by the authors with research on this topic are shown in Figure 5.2.

Trends in the annual research publications on blood supply management for the period 2001–2022 are shown in Figure 5.3.

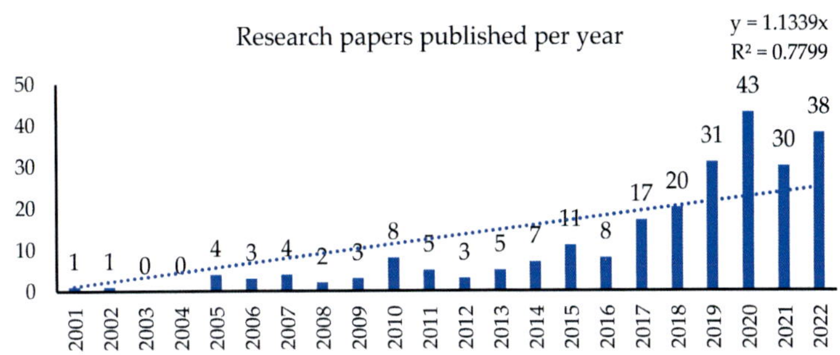

Figure 5.3. Trends on annual research publications on blood supply management for the period 2001–2022

Source: Alghamdi SY. A Review of Blood Delivery for Sustainable Supply Chain Management (BSCM). *Sustainability*. 2023; 15(3):2757. https://doi.org/10.3390/su15032757:

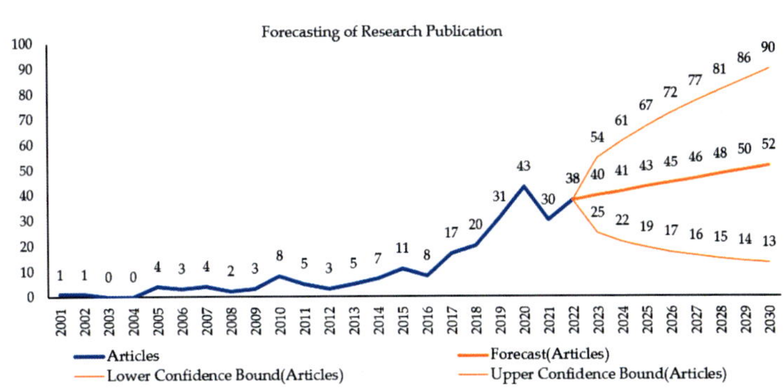

Figure 5.4. Forecasting of research publications.

Source: Alghamdi SY. A Review of Blood Delivery for Sustainable Supply Chain Management (BSCM). *Sustainability*. 2023; 15(3):2757. https://doi.org/10.3390/su15032757:

Trends in the potential annual research publications forecasted for blood supply management for 2021–2030 are shown in Figure 5.4.

Analysis of the leading institutions working on blood supply management are shown in Figure 5.5.

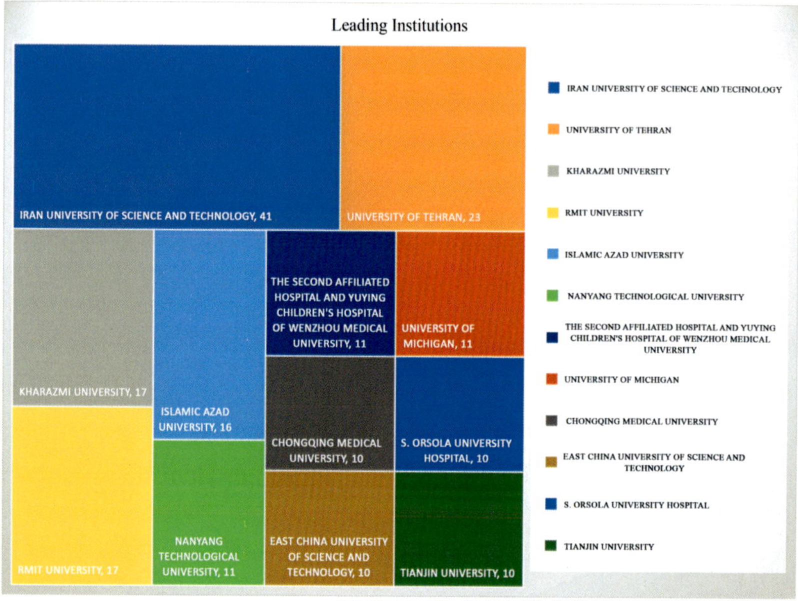

Figure 5.5. The leading institutions working on blood supply management.

Source: Alghamdi SY. A Review of Blood Delivery for Sustainable Supply Chain Management (BSCM). *Sustainability*. 2023; 15(3):2757. https://doi.org/10.3390/su15032757:

Analysis of the country origins of BSCM-based research papers on blood supply management are shown in Figure 5.6.

Given that traumatic injury accounts for a significant number of deaths globally, blood transfusion at the correct time emerges as crucial. A 2022-published article in the *Journal of Clinical Medicine* by Juárez-Vela and team assayed the epidemiology of transfusion practices in severe trauma patients admitted into Intensive Care Units. The study was a multicenter cross-sectional study in 111 Intensive Care Units in Spain and various aspects of blood transfusion were scrutinized. The patient details as presented by the authors have been listed Tables 5.1 and 5.2 (Juárez-Vela et al., 2022).

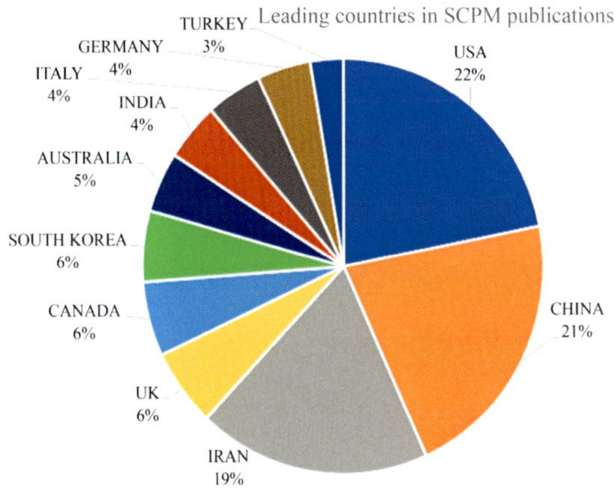

Figure 5.6. Research papers on blood supply management

Source: Alghamdi SY. A Review of Blood Delivery for Sustainable Supply Chain Management (BSCM). *Sustainability*. 2023; 15(3):2757. https://doi.org/10.3390/su15032757:

Table 5.1. Patient details of blood transfusion in the study by Juárez-Vela et al., 2022

		Polytrauma Patients		**Rest of Patients**		***p*-Value**
	(*n* = 109)		**(*n* = 1,330)**			
	Frequency	%	Frequency	%		
Age (mean ± SD)		41.42 ± 18.82		62.61 ± 15.06		<0.001
Gender	Men	93	85.32	858	64.56	<0.001
	Women	16	14.68	471	35.44	
SOFA-Score (Mean ± SD)		5.04 ± 3.46		4.71 ± 3.87		0.406
ICU Stay (Mean ± SD)		13.31 ± 20.53		10.40 ± 17.25		0.092
Mechanical Ventilation	No	45	41.28	714	53.81	0.012
	Yes	64	58.72	613	46.19	
Extrarenal Purification	No	107	98.17	1,209	91.04	0.010
	Yes	2	1.83	119	8.96	

Anticoagulated Patient	No	102	94.44	1,056	79.94	<0.001
	Yes	6	5.56	265	20.06	
Anti-Aggregate Patient	No	101	92.66	1,019	76.79	<0.001
	Yes	8	7.34	308	23.21	
ECMO *	No	107	98.17	1,313	99.09	0.983
	Yes	2	1.83	12	0.91	
Transfusion	No	92	84.40	1,201	90.50	0.041
	Yes	17	15.60	126	9.50	

* Extracorporeal membrane oxygenation.

Source: Juárez-Vela, R., Andrés-Esteban, E. M., Santolalla-Arnedo, I., Ruiz de Viñaspre-Hernández, R., Benito-Puncel, C., Serrano-Lázaro, A., Marcos-Neira, P., López-Fernández, A., Tejada-Garrido, C. I., Sánchez-González, J. L., et al. (2022). Epidemiology and associated factors in transfusion management in intensive care unit. *Journal of Clinical Medicine, 11*(12), 3532. https://doi.org/10.3390/jcm11123532 [© 2022 by the authors. Licensee MDPI, Basel, Switzerland. This article is an open access article distributed under the terms and conditions of the Creative Commons Attribution (CC BY) license: https://creativecommons.org/licenses/by/4.0/].

Table 5.2. Characteristics of transfused patients in the study by Juárez-Vela et al., 2022

Frequency		**Polytrauma Patients**		**Rest of Patients**		***p*-Value**
		%	**Frequency**	**%**		
Age (mean ± SD)		39.00 ± 19.67		66.00 ± 14.37		<0.001
Gender	Men	11	64.71	87	69.05	0.717
	Women	6	35.29	30	30.95	
SOFA-score (mean ± SD)		7.00 ± 3.75		7,00 ± 3.95		0.424
* ICU stay (mean ± SD)		5.00 ± 28.95		5.50 ± 14.37		0.764
Mechanical ventilation	No	3	17.65	44	34.92	0.155
	Yes	14	82.35	82	65.08	
Extrarenal purification	No	16	94.12	100	79.37	0.145
	Yes	1	5.88	26	20.63	
Anticoagulated patient	No	16	94.12	89	72.36	0.052
	Yes	1	5.88	34	27.64	
Anti-aggregate patient	No	13	76.47	101	80.16	0.723
	Yes	4	23.53	25	19.84	
* ECMO	No	17	100.00	121	96.03	0.403
	Yes	0	0.00	5	3.97	

* ICU = Intensive Care Units; ECMO = Extracorporeal membrane oxygenation.

Source: Juárez-Vela, R., Andrés-Esteban, E. M., Santolalla-Arnedo, I., Ruiz de

Viñaspre-Hernández, R., Benito-Puncel, C., Serrano-Lázaro, A., Marcos-Neira, P., López-Fernández, A., Tejada-Garrido, C. I., Sánchez-González, J. L., et al. (2022). Epidemiology and associated factors in transfusion management in intensive care unit. *Journal of Clinical Medicine, 11*(12), 3532. https://doi.org/10.3390/jcm11123532 [© 2022 by the authors. Licensee MDPI, Basel, Switzerland. This article is an open access article distributed under the terms and conditions of the Creative Commons Attribution (CC BY) license: https://creativecommons.org/licenses/by/4.0/].

Table 5.3. The details of transfusions in the study by Juárez-Vela et al., 2022

Transfusion		**Polytrauma Patients**		**Rest of Patients**		***p*-Value**
	(n = 17)		**(n = 126)**			
	Frequency	%	Frequency	%		
Reason for transfusion	Acute anemia with hemodynamic impact	3	17.65%	44	34.92%	0.582
	Acute anemia without hemodynamic impact	3	17.65%	27	21.43%	
	By analytical criteria	7	41.18%	45	35.71%	
	Previous intervention or invasive procedure	2	11.76%	9	7.14%	
	Others	2	11.76%	9	7.14%	
Blood component	Red blood cells concentrate	15	88.24%	113	89.68%	0.736
	Fresh frozen plasma	7	41.18%	48	38.10%	0.940
	Platelets	7	41.18%	50	39.68%	0.848
Number of transfusion events	1	7	41.18%	100	79.37%	0.707
	2	6	35.29%	13	10.32%	
	3	1	5.88%	8	6.35%	
	4	2	11.76%	2	1.59%	
	5 or more	1	5.88%	3	2.38%	

Transfusion control	No	3	17.65%	18	14.29%	0.481
	Yes, in 1 h	1	5.88%	11	8.73%	
	Yes, in 2 h	5	29.41%	16	12.70%	
	Yes, in 3 h	2	11.76%	19	15.08%	
	Yes, in more than 3 h	6	35.29%	62	49.21%	

Source: Juárez-Vela, R., Andrés-Esteban, E. M., Santolalla-Arnedo, I., Ruiz de Viñaspre-Hernández, R., Benito-Puncel, C., Serrano-Lázaro, A., Marcos-Neira, P., López-Fernández, A., Tejada-Garrido, C. I., Sánchez-González, J. L., et al. (2022). Epidemiology and associated factors in transfusion management in intensive care unit. *Journal of Clinical Medicine, 11*(12), 3532. https://doi.org/10.3390/jcm11123532 [© 2022 by the authors. Licensee MDPI, Basel, Switzerland. This article is an open access article distributed under the terms and conditions of the Creative Commons Attribution (CC BY) license: https://creativecommons.org/licenses/by/4.0/].

Of 109 patients, packed red blood cells accounted for the most transfused blood component (93.8%), followed by platelets (43.8%) and whole plasma (37.5%). The details of the transfusion practices as presented by the authors have been provided in Tables 5.3 and 5.4 (Juárez-Vela et al., 2022).

Table 5.4 Blood levels before and after transfusions in the study by Juárez-Vela et al., 2022

Transfused Patient Previous		**Post-Transfusion**			***p*-Value**			
Median		**P25**	**P75**	**Median**	**P25**	**P75**		
Polytrauma patient	hemoglobin	8.6	8.1	10.4	9.65	8.6	10.75	0.077
	MCV	91.2	87.7	95	91.8	87.9	92.5	0.979
	INR	1.13	1	1.2	1.1	1	1.27	0.353
	platelets	151	129	246	183	122	224	0.988
Other patient	hemoglobin	7.6	6.9	8.6	9	8.4	10	<0.001
	MCV	90	86.6	95	90	85.7	93.5	0.004
	INR	1.2	1.08	1.5	1.22	1.09	1.42	0.055
	platelets	148	78	220	137	79	192	0.990

Source: Juárez-Vela, R., Andrés-Esteban, E. M., Santolalla-Arnedo, I., Ruiz de Viñaspre-Hernández, R., Benito-Puncel, C., Serrano-Lázaro, A., Marcos-Neira, P., López-Fernández, A., Tejada-Garrido, C. I., Sánchez-González, J. L., et al. (2022). Epidemiology and associated factors in transfusion management in intensive care unit. *Journal of Clinical Medicine, 11*(12), 3532. https://doi.org/10.3390/jcm11123532 [© 2022 by the authors. Licensee MDPI, Basel, Switzerland. This article is an open access article distributed under the terms and conditions of the Creative Commons Attribution (CC BY) license: https://creativecommons.org/licenses/by/4.0/].

It was concluded that a ratio of red blood cells, platelets, and Fresh Frozen Plasma (FFP) of 2:1:1 is beneficial and outcomes were improved by Massive Transfusion Protocols, especially for trauma patients. Given that auditing blood transfusion emerges as vital in order to prevent wastage, patient blood management (PBM) enters the scene. It is "the appropriate use of blood and blood components, with a goal of minimizing their use" and "the timely application of evidence-based medical and surgical concepts designed to maintain hemoglobin (Hb) concentration, optimize hemostasis and minimize blood loss in an effort to improve patient outcome." Tsante et al. (2021) employed an electronic transfusion request form and a prospective audit of transfusion practice to probe the blood utilization in a Greek tertiary hospital. Statistical analysis revealed that 9,478 RBC units were transfused (mean: 1.75 units per patient) before the PBM as opposed to 9,289 transfused units (mean: 1.57 units per patient) after applying the PBM. However, 21.8% of all the non-bleeding patients received more than one RBC unit transfusion and the restrictive transfusion threshold was not adhered to optimally. These findings serve as a wake-up call for further fine-tuning of relevant PBM approaches. Table 5.5 represents the results published by the researchers (Tsante et al., 2021).

Table 5.5. Hemoglobin Levels of Transfused Patients Without Comorbidities

Study Population	Hb (g/dL)
Total (*n* = 3,099)	7.19 ± 0.79; 7.4 (6.8–7.7)
2nd Department of Internal Medicine (*n* = 1,279)	7.16 ± 0.80; 7.4 (6.8–7.7)
2nd Department of Internal Medicine/ Oncology (*n* = 303)	7.18 ± 0.69; 7.4 (6.9–7.7)
Department of Gastroenterology (n = 5)	7.08 ± 0.46; 6.9 (6.8–7.0)
4th Department of Internal Medicine *(n* = 311)	6.89 ± 1.01; 7.1 (6.4–7.6)
Critical Care Unit *(n* = 118)	7.08 ± 1.06; 7.0 (6.6–7.5)
Hematopoietic Stem Cell Transplantation unit (*n* = 164)	7.54 ± 0.59; 7.6 (7.3–7.9)
Department of Hematology (*n* = 903)	7.29 ± 0.64; 7.4 (7.0–7.7)
Department of Neurology *(n* = 1)	6.7
Emergency Department (*n* = 7)	7.71 ± 1.97; 7.6 (6.6–9.9)
Department of Dermatology (*n* = 4)	7.87 ± 0.20; 7.9 (7.75–8.0)

Department of Cardiology (*n* = 4)	8.27 ± 3.26; 7.65 (7.1–9.45)

Key: Hb, Hemoglobin; RBC, red blood cells. Data are presented as mean ± SD, median, and interquartile range (IQR).

Table 5.6. Transfused Patients Without Comorbidities

Parameters	**Total (*n* = 3099)**
Patients with Hb > 7 g/dL	2,065 (66.6)
Patients with Hb > 8 g/dL	167 (5.3)
RBC (units)/patient	1.18 ± 0.42; 1.0 (1.0–1.0)

Key: Hb, Hemoglobin; RBC, red blood cells. Data are presented as mean ± SD, median, and interquartile range (IQR), or as absolute value (percentage) when appropriate.

Table 5.7. Hemoglobin Levels of Transfused Patients with Comorbidities

Study Population	Hb (g/dL)
Total (*n* = 1,302)	7.62 ± 1.25; 7.5 (7.0–8.1)
2nd Department of Internal Medicine (*n* = 315)	7.37 ± 1.00; 7.5 (7.3–7.6)
2nd Department of Internal Medicine/ Oncology (*n* = 187)	6.72 ± 0.39; 6.8 (6.4–7.1)
Department of Gastroenterology (n = 17)	7.70 ± 0.00; 7.7 (7.7–7.7)
4th Department of Internal Medicine (*n* = 512)	8.37 ± 1.44; 8.2 (7.85–8.90)
Critical Care Unit (*n* = 217)	6.94 ± 0.40; 7.1 (6.6–7.3)
Hematopoietic Stem Cell Transplantation unit (*n* = 5)	7.50 ± 0.00; 7.5 (7.5–7.5)
Emergency Department (*n* = 3)	7.53 ± 0.05; 7.5 (7.5–7.6)
Department of Dermatology (*n* = 1)	7.6
Department of Cardiology (*n* = 34)	7.81 ± 0.83; 7.9 (7.1–8.4)
Cardiac Care Unit (*n* = 11)	7.50 ± 0.00; 7.5 (7.5–7.5)

Key: Hb, Hemoglobin; RBC, red blood cells. Data are presented as mean ± SD, median, and interquartile range (IQR).

5.2. PATIENT BLOOD MANAGEMENT AND THE COVID-19 PANDEMIC

We all still shudder when we think of the COVID-19 pandemic that impacted us on so many levels. To give an example, Gill et al. (2022) reviewed research articles that studied the impact of COVID-19 on behavior at several levels. They have been beautifully summarized by the authors in Table 5.8.

Table 5.8.

Theme	Subtheme	Keyword	Author	Articles	Country	n Sample	Type of Paper
Dietary Behavior	Food security	COVID-19, pandemic, food choices, food security, shopping, purchase, behavior, panic buying, consumer changes, food supply chain, agriculture, food, heart disease, diet, malnutrition, cardiovascular, respiratory, cancer, NCDs, policies, fresh food, fruits, vegetables, immune system, chronic disease, inflammation, global, economy, food crisis, food production, prices, crop, farmer, sales, disruptions, demand, income, trade, poverty, logistics	Bracale, R. et al.	Changes in food choice following restrictive measures due to COVID-19	Italy	10,769 stores	Longitudinal Observational Study
			Di Crosta et al.	Psychological factors and consumer behavior during the COVID-19 pandemic.	Italy	3,833 participants	Cross-sectional Observational Study
			Aday, et al.	Impact of COVID-19 on the Food Supply Chain.	N/A	N/A	Narrative Review

			Branca, et al.	Transforming the food system to fight non-communicable diseases.	N/A	N/A	Editorial
			Glade et al.	Food, nutrition, and the prevention of cancer: a global perspective	Worldwide	N/A	Scoping Review
			He, et al.	Increased consumption of fruit and vegetables is related to a reduced risk of coronary heart disease: Meta-analysis of cohort studies.	United Kingdom	278,459 individuals	Systematic Review
			Luckstead, et al.	Labor issues in the food supply chain amid the COVID-19 pandemic	United States of America	1,648 respondents	Cross-sectional Observational Study
			Bagatini, et al.	Immune system and chronic diseases.	Brazil	N/A	Editorial

			Alam, et al.	Impact of COVID-19 on vegetable supply chain and food security: Empirical evidence from Bangladesh	Bangladesh	120 respondents	Longitudinal Observational Study
			Pan, et al.	The influence of COVID-19 on agricultural economy and emergency mitigation measures in China: A text mining analysis.	China	337 WeChat articles, 490 Weibo articles ~750,000 words	Scoping Review
			Laborde, et al.	COVID-19 risks to Global Food Security	United States of America	N/A	Editorial
			Commitee on World Food Security (CFS) High Level Panel of Experts (HLPE)	Impacts of COVID-19 on food security and nutrition: developing effective policy responses to address the hunger and malnutrition pandemic.	Italy	N/A	Longitudinal Observational Study

			Laborde, et al.	Impacts of COVID-19 on Global Poverty, food security, and diets: Insights from Global Model Scenario Analysis	Worldwide	IFPRI's global general equilibrium model	Cohort Observational Study
	HFP activities	home food procurement (HFP), gardening, home food procurement, COVID-19, pandemic, food insecurity, poverty, foraging, fruits, vegetables, healthy lifestyle, community gardens, social benefit, nutrition, food system, urban agriculture, organic food, community, ecosystem, home gardens, malnourishment, nutritional security	Niles, et al.	Home food procurement impacts food security and diet quality during COVID-19	United States of America	600 residents	Cross-sectional Observational Study
			van den Berg, et al.	Allotment Gardening and health: A comparative survey among allotment gardeners and their neighbors without an allotment	Netherlands	184 respondents (121 + 63 control)	Cross-sectional Observational Study
			Algert, et al.	Vegetable output and cost savings of community gardens in San Jose, California.	United States of America	83 gardeners	Longitudinal Observational Study

			Nova, et al.	Urban Organic Community Gardening to promote environmental sustainability practices and increase fruit, vegetables and organic food consumption.	Portugal	115 city dwellers	Longitudinal Observational Study
			Lampert, et al.	Evidence on the contribution of community gardens to promote physical and mental health and well-being of non-institutionalized individuals: A systematic review	Worldwide	8 studies	Systematic Review
			Lal, et al.	Home Gardening and urban agriculture for advancing food and nutritional security in response to the COVID-19 pandemic.	United States of America	N/A	Editorial

	Alcohol Consumption	alcohol consumption, pandemic, e-commerce, alcohol sales, COVID-19, reward pathway, neural pathways, addiction	Scarpetta, et al. (Organization for Economic Co-operation and Development; OECD)	The effect of COVID-19 on alcohol consumption, and policy responses to prevent harmful alcohol consumption	Worldwide	N/A	Prospective Study
			Koob, et al.	Stress, dysregulation of drug reward pathways, and the transition to drug dependence.	United States of America	N/A	Retrospective Study
			Steffen, et al.	Altered alcohol consumption during COVID-19 pandemic lockdown.	Germany	2,067 participants	Cross-sectional Observational Study
Nutrition	Social behavior	social behavior, anxiety, COVID-19, pandemic, psychological impact, depression, mental health, comfort food, neurophysiological mechanism, intrinsic reward mechanism, hyperpalatable food, obesity, inflammation, visceral weight gain, cardiovascular, adipose tissue, complications, calories, nutrient density, hunger, insulin resistance, cancer, body mass index (BMI), intensive care unit, SARS-Cov-2	Wang, et al.	Immediate psychological responses and associated factors during the initial stage of the 2,019 coronavirus disease (COVID-19) epidemic among the general population in China	China	1,210 respondents	Cross-sectional Observational Study

			Weltens, et al.	Where is the comfort in Comfort Foods? mechanisms linking fat signaling, reward, and emotion	N/A	N/A	Narrative Review
			Fazzino, et al.	Hyper-palatable foods: Development of a quantitative definition and application to the US Food System Database	United States of America	75 HPF descriptors	Systematic Review
			Hauner,	Secretory factors from human adipose tissue and their functional role: Proceedings of the nutrition society	N/A	N/A	Editorial
			Fuhrman, et al.	Changing perceptions of hunger on a high nutrient density diet	United States of America	768 participants	Cross-sectional Observational Study

			Shuster, et al.	The clinical importance of visceral adiposity: A critical review of methods for visceral adipose tissue analysis.	Worldwide	N/A	Narrative Review
			Mohammad, et al.	Obesity and COVID-19: What makes obese host so vulnerable?	Worldwide	N/A	Narrative Review
			Simonnet, et al.	High prevalence of obesity in severe acute respiratory syndrome coronavirus-2 (SARS-CoV-2) requiring invasive mechanical ventilation.	France	124 patients	Retrospective Study
			Ouchi, et al.	Adipokines in inflammation and metabolic disease.	Worldwide	N/A	Narrative Review

	Mediterranean diet	Mediterranean diet, health benefits, COVID-19, adipose tissue, inflammation, olive oil, polyphenols, cardiovascular health, anti-inflammatory, thrombosis, mast cells, polyphenols, immunomodulation, cytokines, diabetes, chronic disease, atherosclerosis, food choice, mortality, financial cost, hospitalization	Bach-Faig, et al.	Mediterranean diet pyramid today. science and cultural updates	Mediterranean region	N/A	Prospective Study
			Angelidi et al.	Mediterranean diet as a nutritional approach for COVID-19.	N/A	N/A	Editorial
			Theoharides, et al.	Coronavirus 2019, Microthr-omboses, and platelet activating factor	N/A	N/A	Editorial
			Shakoor, et al.	Immuno-modulatory effects of dietary polyphenols.	Worldwide	167 papers	Systematic Review
			Ding, et al.	Regulation of immune function by polyphenols	N/A	N/A	Narrative Review

			Maiorino, et al.	Mediterranean diet and COVID-19: Hypothesizing potential benefits in people with diabetes	N/A	N/A	Editorial
			Casas, et al.	The immune protective effect of the Mediterranean diet against chronic low-grade inflammatory diseases	N/A	N/A	Narrative Review
			Greene, et al.	Negative association between Mediterranean diet adherence and COVID-19 cases and related deaths in Spain and 23 OECD countries: An ecological study	OECD countries	24 countries	Cohort Observational Study

			Lampropoulos, et al.	Effects of Mediterranean diet on hospital length of stay, medical expenses, and mortality in elderly, hospitalized patients: A 2-year observational study	Greece	183 patients	Longitudinal Observational Study
			Mohajeri, et al.	The food choice determinants and adherence to Mediterranean diet in Iranian adults before and during COVID-19 lockdown: Population-based study	Iran	2,540 adults	Cross-sectional Observational Study
	Intermittent Fasting	glucose, glycolysis, COVID-19, cytokine storm, intermittent fasting, metabolism, metabolic pathways, respiratory virus, SARS-CoV-2, diabetes, pandemic	Wang, et al.	O-glcnac transferase promotes influenza A virus–induced cytokine storm by targeting interferon regulatory factor–5.	N/A	N/A	Experimental Study

			Yang, et al.	The signal pathways and treatment of cytokine storm in COVID-19	N/A	N/A	Scoping Review
			Codo, et al.	Elevated glucose levels favor SARS-CoV-2 infection and monocyte response through a HIF-1α/ glycolysis-dependent axis	N/A	N/A	Experimental Study
			Lee, et al.	Intermittent fasting: Physiological implications on outcomes in mice and men	N/A	N/A	Narrative Review
			Albosta, et al.	Intermittent fasting: Is there a role in the treatment of diabetes? A review of the literature and guide for Primary Care Physicians.	United States of America	17 articles	Narrative Review
			Ealey, et al.	COVID-19 and obesity: Fighting two pandemics with intermittent fasting	N/A	N/A	Narrative Review

Food Options	Food Delivery	food delivery, food delivery apps, FDAs, online transactions, COVID-19, online food purchase lockdown, economic impacts, takeaways, diet quality, food choice, obesity, weight gain, eye level, psychology, supermarket purchase, nutrition, inflammation	Li, et al.	Review of online food delivery platforms and their impacts on sustainability	N/A	N/A	Narrative Review
			Muangmee, et al.	Factors determining the behavioral intention of using food delivery apps during COVID-19 pandemics	Thailand	402 respondents	Cross-sectional Observational Study
			Duda-Chodal, et al.	COVID-19 pandemic and food: Present knowledge, risks, consumers fears and safety.	Poland	N/A	Scoping Review
			Stephens, et al.	Food delivery apps and the negative health impacts for Americans	United States of America	N/A	Editorial

			Wang, et al.	Hunger for Home Delivery: Cross-sectional analysis of the nutritional quality of complete menus on an online food delivery platform in Australia.	Australia	13,841 food items from 202 outlets	Cross-sectional Observational Study
Physical Activity and Sedentary Behaviors	Physical Activity; Sedentary Behaviors	physical activity, sedentary behavior, energy expenditure, factors affecting, COVID-19, isolation, leisure time, work from home (WFH)	Tremblay, et al.	Physiological and health implications of a sedentary lifestyle.	Canada	N/A	Narrative Review
			Füzéki, et al.	Physical activity During COVID-19 INDUCED lockdown: Recomm-endations.	Germany	N/A	Editorial
			Tison, et al.	Worldwide effect of COVID-19 on physical activity: A descriptive study.	187 countries	455,404	Longitudinal Observational Study
			Genin, et al.	Effect of a 5-month worksite physical activity program on tertiary employee's overall health and fitness.	United States of America	95 employees	Longitudinal Observational Study

			Bourdas, et al.	Impact of COVID-19 lockdown on physical activity in a sample of Greek adults.	Greece	8,495	Cross-sectional Observational Study
			Tan, et al.	Physical activity, Sedentary behavior, and Weight status of university students during the COVID-19 Lockdown: A Cross-National comparative study.	Malaysia	254 students	Cross-sectional Observational Study
			Constandt, et al.	Exercising in times OF Lockdown: An analysis of the impact of COVID-19 on levels and patterns of exercise among adults in Belgium.	Belgium	13,515	Cross-sectional Observational Study

Effects of Restrictive Measures towards Dietary Habits and Physical Activity	Dietary Habits; Physical Activity	diet, dietary habits, eating behavior, lifestyle, physical health, physical activity, sedentary behavior, pandemic, confinement, countries, changes, comparisons, quarantine, COVID-19	Shimpo, et al.	Shimpo, et al., Factors associated with dietary change since the outbreak of COVID-19 in Japan.	Japan	6,000	Cross-sectional Observational Study
			Wang, et al.	Bidirectional influence of the COVID-19 pandemic lockdowns on health behaviors and quality of life among Chinese adults.	China	2,289	Cross-sectional Observational Study
			Rodríguez-Pérez et al.	Changes in dietary behaviors during the COVID-19 outbreak confinement in the Spanish COVIDiet study	Spain	7,514 participants	Cross-sectional Observational Study

			Papandreou, et al.	Comparing eating behaviors, and symptoms of depression and anxiety between Spain and Greece during the COVID -19 outbreak: Cross-sectional analysis of two different confinement strategies.	Spain, Greece	1,841 total (1,002 in Spain, 839 in Greece)	Cross-sectional Observational Study
			Sañudo, et al.	Objectively-assessed physical activity, sedentary behavior, smartphone use, and sleep patterns pre- and during-COVID-19 quarantine in young adults from Spain.	Spain	22 students	Longitudinal Observational Study

Source: Gill et al. (2022) [].

5.3. THE MAIN ASPECT IN THE CONTEXT OF BLOOD MANAGEMENT

According to a research article in JAMA, the American Red Cross announced its "worst crisis in a decade." This becomes pertinent as it supplies 40% of blood in the US (Rubin, 2022).

Below is a statement from the Red Cross (January 11, 2022) showing the predicament:

"*The American Red Cross is facing a national blood crisis – its worst blood shortage in over a decade, posing a concerning risk to patient care. Amid this crisis, doctors have been forced to make difficult decisions about who receives blood transfusions and who will need to wait until more products become available. Blood and platelet donations are critically needed to help prevent further delays in vital medical treatments.*"

"While some types of medical care can wait, others can't," said Dr. Pampee Young, chief medical officer of the Red Cross. "Hospitals are still seeing accident victims, cancer patients, those with blood disorders like sickle cell disease, and individuals who are seriously ill who all need blood transfusions to live even as Omicron cases surge across the country. We're doing everything we can to increase blood donations to ensure every patient can receive medical treatments without delay, but *we cannot do it without more donors. We need the help of the American people."*

—Red Cross Press Release January 11, 2022

Let us see another statement from a research paper that further illuminated the grim situation:

"Although there is a paucity of published evidence to date analyzing the real-world impact of the blood shortage, there certainly was an impact on day-to-day clinical care. Patients who would otherwise be transfused for acute anemia due to trauma or chemotherapy were not transfused, surgeries were canceled due to the lack of emergency blood products, and patients with sickle cell disease were unable to have their routine monthly blood transfusions or erythrocytapheresis to prevent stroke or other severe complications."

—McGann PT, Weyand AC. Lessons learned from the COVID-19 pandemic blood supply crisis. J Hosp Med. 2022;17(7):574–576. doi:10.1002/jhm.12843

Table 5.9. The Trend of Blood Donations in Spain (* Prepared from the weekly reports of the Ministry of Health)

	9–15 March	16–22 March	23–29 March	30 March–3 April	6–12 April	13–19 April	20–26 April	27 April–3 May	4–10 May	11–17 May	**18–24 May**	**25–31 May**	**1–7 June**	**8–14 June**	**15–21 June**	**22–28 June**
Week	11	12	13	14	15	16	17	18	19	20	21	22	23	24	25	26
Donation	33,760	30,151	22,595	21,899	20,496	22,395	26,541	24,688	27,480	32,997	32,748	32,174	31,562	32,589	35,239	28,085
Difference	2880	−729	−8285	−8981	−10,384	−8485	−4339	−6192	−3400	2117	1868	1294	682	1709	4359	−2795
(%)	9.3	−2.4	−26.8	−29.1	−33.6	−27.5	−14.1	−20.1	−11.0	+6.9	+6.1	+4.2	+2.2	+5.5	+14.1	−9.1

Source: García-Erce, et al. (2021). Blood Donations and Transfusions during the COVID-19 Pandemic in Spain: Impact According to Autonomous Communities and Hospitals. *International Journal of Environmental Research and Public Health.* 18(7), 3480:

The first few weeks of the pandemic saw a dip in blood donations from 20% to 67% in Washington State, China, India, The Netherlands, Saudi Arabia, Iran, and southern Italy. The major causes attributed included overall illness in the population and fewer numbers of medical personnel which caused a dip in transfusions. Further, another factor was the lack of sensitive and rapid testing facilities at that time. In Spain, compared to the donation status in 2018, blood transfusions dipped by 20% between week 11 and week 25 after the declaration of the State of Emergency.

With this background, García-Erce et al. (2021) investigated how blood transfusion and distribution were impacted by the pandemic at regional transfusion centers (RTCs), representing 6.12% of the Spanish population. The consumption of packed RBCs, plasma, and platelets showed a decrease of 24.5%, 45.3%, and 25.3%, respectively. Furthermore, the RBC consumption was identified as significantly lowered. The findings of the authors have been presented below for increased comprehension.

Table 5.10. The Trend of Blood Donations Grouped by Communities and the Time Frames in Spain During the Pandemic

Region		January and February	1–15 March	16 March–10 May	11–24 May	25 May–8 June	11 May–8 June	16 March–8 June
Navarra	2020	3,581	800	2,294	773	983	1,756	4,050
	Mean 18–19	3,803.5	976	3,396.5	867	910.5	1,777.5	5,174
	Difference (%)	−5.8	−18.0	−32.5	−10.8	8.0	0.1	−21.7
Aragón	2020	7,155	1,816	5,536	1,668	1,889	3,557	9,093
	Mean 18–19	7,170	1,636	5,861.5	1,723.5	1,879.5	3,603	9,464.5
	Difference (%)	−0.21	11.0	−5.6	−3.2	−0.5	−1.3	−3.9
La Rioja	2020	1,663	397	1,522	394	470	864	2,386
	Mean 18–19	1,637	447.5	1,598	410	405.5	815.5	2,413.5
	Difference (%)	1.6	−11.3	−4.8	−3.9	+15.9	+5.9	−1.1
Cantabria	2020	4,139	642	2,018	818	793	1,611	3,629
	Mean 18–19	3,752.5	905	3,245	810.5	837.5	1,648	4,893
	Difference (%)	10.3	−29.1	−37.8	+0.9	−5.3	−2.2	−25.8
Total	Average percentage change	1.5 (−9.2, 12.1)	−11.8 (−38.7, 15.1)	−20.1 (−47.9, 7.6)	−4.3 (−12.0, 3.5)	4.8 (−9.9, 19.4)	0.6 (−5.2, 6.4)	−13.1 (−32.9, 6.7)
	p-value	0.693	0.256	0.104	0.179	0.377	0.754	0.125

Table 5.11. Consumption of Red Blood Cells, Plasma and Platelets in Hospital Centers vs. That of the Same Period of the Previous Biennium

		1 January–28 February	1–15 March	16 March–10 May	11–24 May	25 May–8 June
Red Blood Cells	Average (%)	−3.6 (−11.0, 3.9)	−17.6 (−28.2, −7.1)	−24.5 (−30.0, −19.1)	−5.8 (−21.9, 10.3)	0.6 (−11.5, 12.7)
	p-value	0.307	0.004	<0.001	0.437	0.915
Platelets	Average (%)	2.4 (−25.9, 29.8)	−2.65 (−83.2, 77.9)	−25.3 (−36.5, −14.1)	−16.9 (−37.5, 3.7)	−22.9 (−36.4, −9.4)
	p-value	0.846	0.942	0.001	0.097	0.005
Plasma	Average (%)	−15.6 (−53.7, 22.5)	16.0 (−59.4, 91.3)	−45.3 (−76.7, −14.6)	−9.1 (−49.1, 30.8)	−32.9 (−74.1, 8.3)
	p-value	0.379	0.638	0.009	0.618	0.103

Table 5.12. Changes in the Concentrated Red Blood Cell Consumption by the Hospital

		1 January–28 February	1–15 de March	16 March–10 May	11–24 May	25 May–8 June
Navarra	2020	2,410.0	554.0	1,502.0	717.0	510.0
	Mean 18–19	2,543.5	696.5	2,405.5	1,080.0	720.5
	Difference (%)	−5.25	−20.46	−36.91	−33.61	−29.22
Aragón	2020	2,399.00	581.00	1,673.00	498.00	597.00
	Mean 18–19	2,365.00	628.00	2,193.00	539.00	568.00
	Difference (%)	1.44	-7.48	-23.71	-7.61	5.11
La Rioja	2020	1,521.00	305.00	1,157.00	371.00	397.00
	Mean 18–19	1,545.50	394.00	1,440.50	288.50	368.50
	Difference (%)	−1.59	−22.59	−19.68	28.60	7.73
Valdecilla	2020	2,303.0	463.0	1,642.0	597.0	507.0
	Mean 18–19	2,499.0	622.5	2,095.0	557.5	580.5
	Difference (%)	−7.84	−25.62	−21.62	7.09	−12.66
Parc Tauli	2020	1,086	249.00	773.00	224.00	–
	Mean 18–19	1,000	240.50	944.50	263.00	–
	Difference (%)	8.6	3.53	−18.16	−14.83	–
San Joan Despi	2020	626	175.00	489.00	202.00	171.0
	Mean 18–19	832.0	200.00	747.00	152.50	167.5
	Difference (%)	−24.76	−12.50	−34.54	32.46	2.09
La Paz	2020	3,604.0	714.00	2,634.00	667.00	794.0
	Mean 18–19	3,346.0	1,321.00	3,074.50	752.50	723.5
	Difference (%)	7.71	−45.95	−14.33	−11.36	9.74
Getafe	2020	878.0	197.0	620.0	192.0	193.0
	Mean 18–19	1,022.0	282.5	926.5	258.5	215.0
	Difference (%)	−14.09	−30.27	−33.08	−25.73	−10.23

Puerta del Hierro	2020	2,327.0	483.0	1,539.0	499.0	598.0
	Mean 18–19	2,188.5	572.0	1,923.0	528.5	479.0
	Difference (%)	6.33	−15.56	−19.97	−5.58	24.84
Infanta Leonor	2020	881.0	209.0	547.0	145.0	201.0
	Mean 18–19	933.5	208.0	715.0	199.5	186.5
	Difference (%)	−5.62	0.48	−23.50	−27.32	7.77
Total	Average percentage change	−3.6 (−11.0, 3.9)	−17.6 (−28.2, −7.1)	−24.5 (−30.0, −19.1)	−5.8 (−21.9, 10.3)	0.6 (−11.5, 12.7)
	p-value	0.307	0.004	<0.001	0.437	0.915

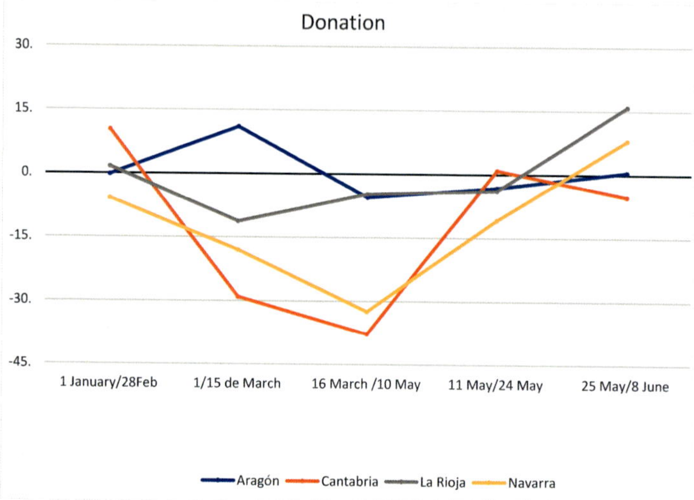

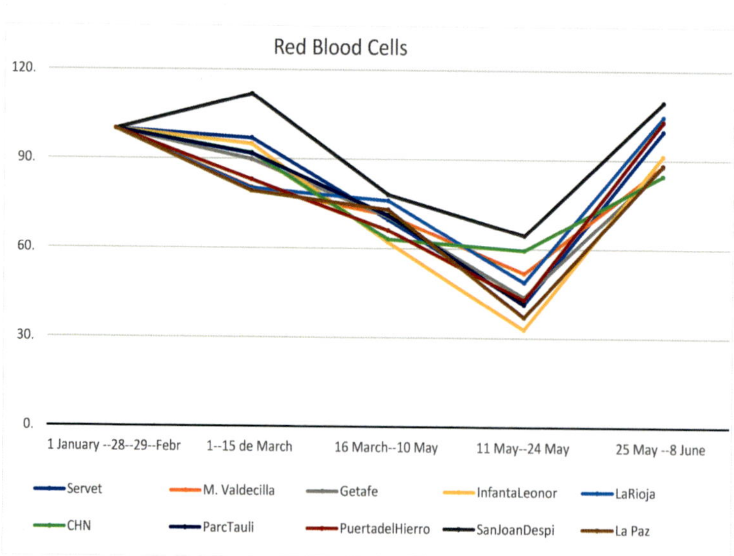

(**A**) Red blood cells concentrate

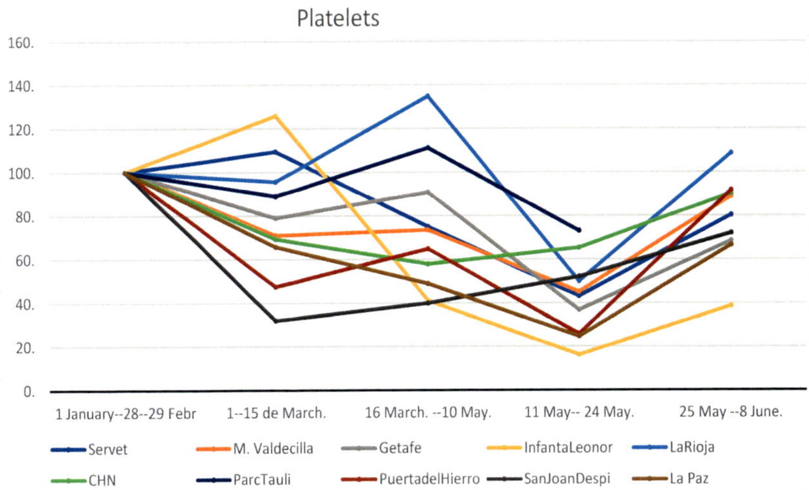

(**B**) Platelets

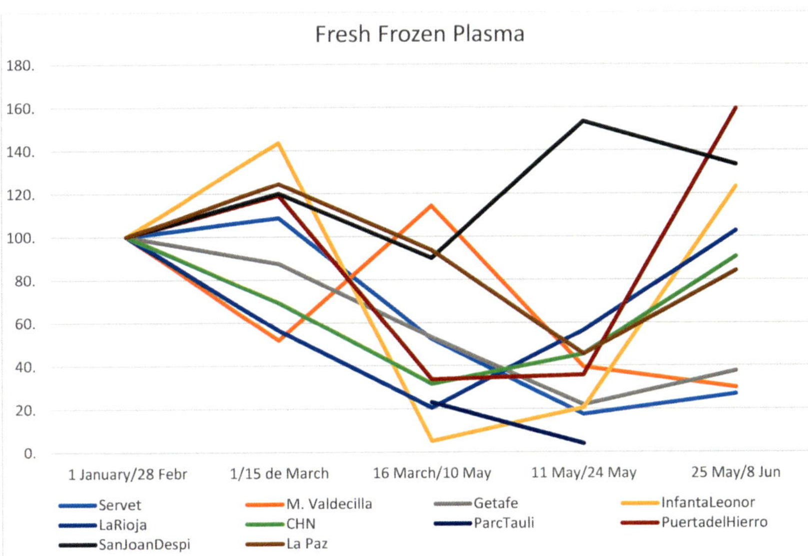

(**C**) Fresh frozen plasma

Figure 5.7. Trends in blood donation during the COVID-19 Pandemic in Spain

García-Erce, et al. (2021). Blood Donations and Transfusions during the COVID-19 Pandemic in Spain: Impact According to Autonomous Communities and Hospitals. *International Journal of Environmental Research and Public Health.* 18(7), 3480:

An important finding that emerges here is a lack of common and uniform criteria among hospitals or the autonomous communities. Further fine-tuning of this process is necessitated along with planned relevant and transparent strategies for efficient PBM approaches (García-Erce et al, 2021).

5.4. CASE STUDY 1

A retrospective cross-sectional study published in *Thalassemia Reports* probed the status of transfusion and control in transfusion-dependent beta-thalassemia (TDT) patients (n=101) at the Mangunkusumo Hospital (CMH) in Jakarta, Indonesia from 2019–2021. This becomes pertinent given that when patients have low pre-transfusion Hb levels, several complications emerge including a low quality of life, pulmonary hypertension, thrombosis, and hepatic iron overload. Blood transfusion becomes vital in this scenario to boost the Hb level to normal values. However, the pandemic-associated blood shortage created a vicious cycle of a lack of blood volume and the unmet gap. Figure 5.8 illustrates the condition.

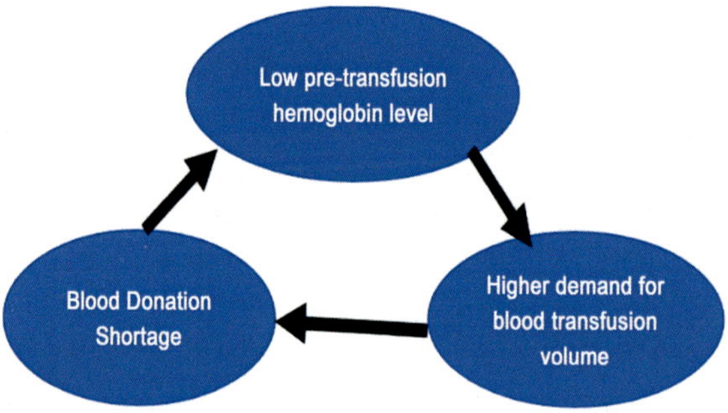

Figure 5.8. A vicious cycle of a lack of blood volume and the unmet gap during the COVID-19 pandemic

Source: Rahmartani LD, Dewi MK, Iskandar SD, Pratanata AM, Ilmana G, Sari TT, Lubis AM, Wahidiyat PA. Impact of COVID-19 Pandemic on Pre-Transfusion Hemoglobin Level and Frequency of Transfusion in Transfusion-Dependent Thalassemia Patients in Indonesia. *Thalassemia Reports*. 2023; 13(1):1-9. https://doi.org/10.3390/thalassrep13010001

Sadly, the researchers found that 59.4% of subjects had suboptimal Hb levels of <9.0 g/dL before the pandemic, which rose to 71.3% during the pandemic. The poor control of the Hb levels highlights the gap in blood availability, potentially creating the probability of complications. Tables 5.13–5.16 compiled by the researchers show the results of the analysis (Rahmartani et al., 2023).

Table 5.13. The Demographic Characteristics of the Study

Characteristic	***n* = 101**	**%**
Gender		
Male	51	50.5
Female	50	49.5
Age Group (Years Old)		
2–18	47	46.5
>18	54	53.5
Occupation		
Student/Unemployed	74	73.3
Employee	14	13.8
Entrepreneur	9	8.9
Housewife	2	2.0
Freelancer	2	2.0
Diagnosis		
β-Thalassemia	84	83.2
β-HbE Thalassemia	17	16.8
Transfusion Interval		
Every 2 weeks	50	49.5
Every 3 weeks	34	33.7
Every 4 weeks	17	16.8

Source: Rahmartani LD, Dewi MK, Iskandar SD, Pratanata AM, Ilmana G, Sari TT, Lubis AM, Wahidiyat PA. Impact of COVID-19 Pandemic on Pre-Transfusion Hemoglobin Level and Frequency of Transfusion in Transfusion-Dependent Thalassemia Patients in Indonesia. *Thalassemia Reports*. 2023; 13(1):1-9. https://doi.org/10.3390/thalassrep13010001

Table 5.14. Pre-Transfusion Hb Level of TDT Patients and the Pandemic

Hb Level	**Optimal (%)**	**Suboptimal (%)**
Pre-pandemic	41 (40.6)	60 (59.4)
During pandemic	29 (28.7)	72 (71.3)
Average Hb level change	*n* = 101	%
Decreased average Hb level	69	68.3
No change in average Hb level	3	3.0
Increased average Hb level	29	28.7

Source: Rahmartani LD, Dewi MK, Iskandar SD, Pratanata AM, Ilmana G, Sari

TT, Lubis AM, Wahidiyat PA. Impact of COVID-19 Pandemic on Pre-Transfusion Hemoglobin Level and Frequency of Transfusion in Transfusion-Dependent Thalassemia Patients in Indonesia. *Thalassemia Reports*. 2023; 13(1):1-9. https://doi.org/10.3390/thalassrep13010001

Table 5.15. Pre-Transfusion Hemoglobin Level before and During the Pandemic

Hb Level	**Mean (Min–Max) (g/dL)**	***p*-Value**
Pre-pandemic	8.70 (5.94–11.08)	<0.001
Post-pandemic	8.45 (3.48–10.18)	–

Source: Rahmartani LD, Dewi MK, Iskandar SD, Pratanata AM, Ilmana G, Sari TT, Lubis AM, Wahidiyat PA. Impact of COVID-19 Pandemic on Pre-Transfusion Hemoglobin Level and Frequency of Transfusion in Transfusion-Dependent Thalassemia Patients in Indonesia. *Thalassemia Reports*. 2023; 13(1):1-9. https://doi.org/10.3390/thalassrep13010001

Researchers Vasiljević-Jovanović and team (2021) scrutinized the usage of blood products in patients hospitalized due to COVID-19 at the University Hospital Medical Center BezanijskaKosa in Belgrade, Serbia. The analysis of 90 patients revealed that 304 (139 in ICU, 165 in other wards) red blood cell concentrates (RBC) with a mean of 3 units/patient (range 1–14) were transfused over the second wave. Interestingly, higher usage of RBCs could be predicted by comorbidities, illness severity, and stay in the ICU. This paves the way for further analysis of using these factors to stock blood and be prepared, especially given that we remember how the pandemic severely tested the healthcare industry.

The findings of the study have been presented below:

Transfusion data of the patients hospitalized due to COVID-19 at the University Hospital Medical Center BezanijskaKosa in Belgrade, Serbia Vasiljević-Jovanović B, Milenković M, Mijović L, Bukumirić Z, Šantrić-Milićević M, Božanić M, Milutinović V, Zdravković M. Evaluation of the Use of Blood Products in ICU Hospitalized COVID-19 Patients. *Surgeries*. 2021;

Table 5.16. Baseline transfusion data

Variables	All Patients n = 90
Age, median (range)	72 (23–95)
Gender, male/female, n (%)	50 (55.6)/40 (44.4)
Pts in ICU/ward, *n* (%)	36 (40)/54 (60)
Hospital days, median (range)	23 (3–73)
Hospital days in ICU, median (range)	9 (0–73)
Invasive respiratory support days, median (range)	0.5 (0–72)
Chest CT severity score, median (range)	20 (4–25)
All comorbidities, *n* (%)	86 (95.6)
Cardiovascular/HTA, *n* (%)	71 (78.9%)
Neurological/neuromuscular	17 (18.9%)
DM, *n (*%)	24 (26.7%)
COPD/asthma, *n* (%)	11 (12.2%)
Immunodeficiency/HIV, *n* (%)	6 (6.7%)
Cancer, *n* (%)	18 (20.0%)
CLD, *n* (%)	3 (3.3%)
CKD, *n* (%)	18 (20.0%)
Anemic pts. at admission, *n* (%)	50/90 (55.6)
Transfused RBC/Plt/FFP/Cryo, *n*	304/42/96/58
Transfused RBC ICU/ward in total, *n* (%)	139 (45.7%)/165 (54.3%)
Transfused Units Per Patient, Median (range):	
RBC	3 (1–14)
Plt	1 (1–7)
FFP	3.5 (1–16)
Cryo	10 (6–12)
Hemoglobin at transfusion, g/dL ICU/ward, median (range)	75 (59–121)/77 (53–96)
In-hospital mortality n (%)	36 (40.0%)

Key: categorical variables are expressed as numbers and percentages; continuous variables are expressed as the mean (±standard deviation) or median (IQR). ICU: intensive care unit; RBC: red blood cells; Plt: platelets; FFP: fresh frozen plasma; Cryo: cryoprecipitate.

Regression analysis of the number of RBC doses as a dependent variable, N = 90

Table 5.17. Regression analyses of the data

Variable	**Univariate**		**Multivariate**	
	B	***p*-Value**	**B**	***p*-Value**
Gender (m vs. f)	<0.001	1.000		
Age (years)	<0.001	1.000		
Total hospital days	0.059	0.010	0.041	0.071
Hospital days in the ICU	0.054	0.023		
Respiratory support days	0.047	0.093		
Place of residence (Belgrade vs. others)	<0.001	1.000		
Ward (ICU vs. others)	2.000	0.002	−0.143	0.874
Chest CT severity score	−0.050	0.570		
Comorbidities	1.000	0.002	0.847	0.009
Anticoagulant therapy	−1.000	0.434		
Illness severity (moderately severe vs. severe)	2.000	0.001	1.296	0.173

5.5. CASE STUDY 2

In another study of the relevance of PBM with regard to the COVID-19 pandemic, the complete record of COVID-19 patients from March 2020 to August 2020 was analyzed (inclusive of blood transfusion data) in the Blood Bank RLJ Hospital Department of Pathology attached to Sri Devaraj Urs Medical College & Hospital, India. 35% of the 200 COVID-19 positive cases studied received packed Red Blood Cells (PBRBCs), of which 64.28% were associated with severe anemia and pregnancy-related anemia. Further, a platelet transfusion attributed to comorbidities was recorded in 18 patients (25.7%), and 7 (10%) received Fresh Frozen Plasma (FFP) (Hadimani et al., 2022). The management approach employed (The Cleveland Clinic Blood Management) and the PBM strategies put forth by the researchers have been presented below (Hadimani, et al., 2022).

Adopted from Cleveland Clinic blood management

Hemoglobin	Symptoms	Action taken
Hb: < 7 gm	Active bleeding	Transfuse PRBC to achieve Hb between 8 - 9 gm%
	Symptomatic	Transfuse 1 unit PRBC & reassess the symptoms
	Asymptomatic	Iron studies & ferritin levels for all patients & treat the iron deficiency anemia by IV Iron
Hb: > 7 gm	Active bleeding	Transfuse PRBC to achieve Hb between 8 - 9 gm%
	Symptomatic	Transfuse 1 unit PRBC & reassess the symptoms. If symptomatic then transfuse 1 more unit of PRBC then Iron studies & ferritin levels, treat with IV iron if iron deficiency anemia is present
	Asymptomatic	Medical management by assessing the iron, ferritin, Vitamin B12 levels. IV iron & IV vitamin B12 if iron or vitamin B12 deficiency anemia is present

PBM Strategies

	Considerations	Possible actions
Red blood cell Usage	Red blood cell shortages	Patients with mild anemia should be given oral and IV Iron sucrose and can reduce the usage of PRBCs.
Platelet usage	Platelet shortages for prophylactic transfusion	Prophylactic usage of platelets should be minimized in patients with thrombocytopenia without any bleeding symptoms and careful monitoring of patients to be done.
Major bleeding	Blood shortages for patients with bleeding	Patients with bleeding should be transfused 1:1:1 for red blood cells, plasma, and platelets, or 1:1 for red blood cells and plasma if platelets are not available. If there is shortage of red blood cells consider giving plasma first or blood components at ratios of 1:2:1
Alternatives for Transfusion	At times of severe shortage alternatives to transfusion can be given.	Use of, Erythropoitein, oral or IV Iron sucrose & Tranxemic acid for bleeding.

Importantly, the authors corroborated the importance of PBM, especially in sailing through the shortage of blood during the COVID-19 pandemic. Thus, studies encompassing regulatory policies and management are warranted in formulating a blood management plan to handle any emergency.

5.6. IMPLEMENTING PATIENT BLOOD MANAGEMENT

According to the WHO Action framework to advance universal access to safe, effective, and quality-assured blood products, 2020–2023; in 2020, patient blood management is "a set of evidence-based practices to optimize medical and surgical patient outcomes through preservation of the patient's own blood." Despite the initiatives, there are several barriers such as transfusion behavior. For example, the rates of blood transfusion varied from 7.8% to 92.8% for RBCs, 0% to 97.5% for fresh-frozen plasma, and 0.4% to 90.4% for platelets in a hospital-level variation study in the use of allogeneic red blood cell (RBC), fresh-frozen plasma, and platelet transfusions in patients undergoing coronary artery bypass graft (CABG) surgery (Bennett-Guerrero et al, 2010). Another aspect is the attitude of physicians as evidenced in an early face-to-face survey of 122 general surgeons, orthopedic surgeons, and anesthesiologists in three hospitals regarding transfusion decision-making. Sadly, each transfusion risk was estimated correctly by less than 50% of the physicians surveyed, and only 31% responded correctly to a set of four questions concerning transfusion indications (Salem-Schatz et al, 1990). A 2021-published review by the PBM Implementation Group of 1–4 PBM implementers from 12 countries in Asia, Latin America, Australia, Central and Eastern Europe, the Middle East, and Africa amalgamated various aspects of PBM in terms of drivers, barriers, measures, and stakeholders. Salient reports and findings as outlined and presented by the authors have been presented below.

The implementation matrix for patient blood management developed by the researchers (Hofmann et al. 2021).

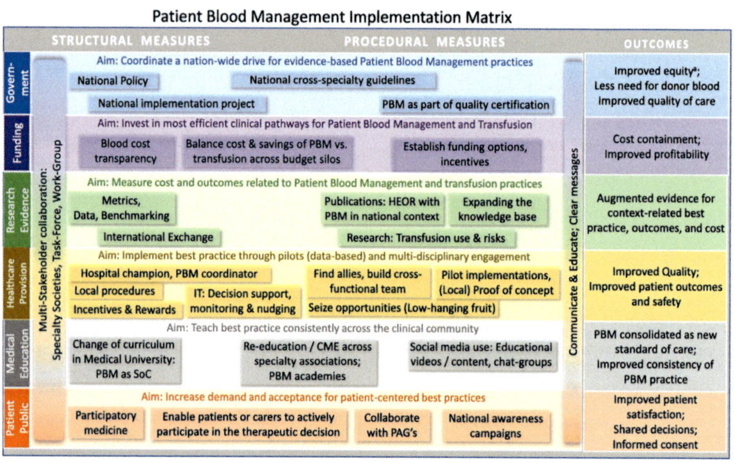

Figure 5.9. The current status and approach to implementing patient blood management at the national or local levels.

Source: Hofmann et al. (2021).

Table 5.18 below outlines a summary of responses to the interviews conducted by the study in from 12 countries in Asia, Latin America, Australia, Central and Eastern Europe, the Middle East, and Africa. Here, 'Mature' describes a high level of implementation; 'Advanced' denotes a strategic, coordinated approach towards general implementation; 'Initiation' describes the occurrence of few individual initiatives with a low level of coordination.

Key: BB Blood Bank, EU European Union, MoH Ministry of Health, N.A. Not available, NBA National Blood Authority, PBM Patient Blood Management

Table 5.18. A summary of responses to the interviews conducted by the study from 12 countries in Asia, Latin America, Australia, Central and Eastern Europe, the Middle East, and Africa

	Australia	Brazil	China	Croatia	Greece	Lebanon	Mexico	Saudi Arabia	South Africa	South Korea	Switzerland	Turkey
	AUS	**BRA**	**CHN**	**HRV**	**GRE**	**LBN**	**MEX**	**SAU**	**ZAF**	**KOR**	**CHE**	**TUR**
Stage	MATURE; National Policy; Broad uptake	INITIATION; Scattered individual leadership	ADVANCED; National Drive	INITIATION; Specialist society drive	INITIATION; National drive through BB	INITIATION; individual initiatives	INITIATION; national project and coordinator	INITIATION; in some hospitals / institutions	INITIATION; individual initiatives.	ADVANCED; implementa-tion project; strong PBM society	INITIATION / ADVANCED; in some hospitals / institutions	ADVANCED; Broad awareness through cross-specialty implementation project
Approach												
Top Down	NBA; States Legal framework, standards & guidance Quality certification	N.A.	National Health Commission, PBM as part of blood supply & demand strategy	MoH informed Some coordination by specialist society (network)	National BB (with MoH) and in cooperation with hospital managers	Original political drive suspended due to political situation	National project with Quality certification	National interest group, but not much coordination	N.A.	National drive: collaboration program for pilots, data collection, and setting standards	Limited due to strong federal healthcare policy structure	EU Project for PBM implement-tation
Bottom up	Initially driven by individual leadership and multi-disciplinary teams in hospitals	Individual Leadership	Clinical champions in leading pilot hospitals	Clinical/ de-partmental champions	Individual initiatives in some hospitals / institutions	Individual initiatives in some hospitals / institutions	Coordinated bottom up	Introduced in some hospitals / institutions	Individual initiatives	Individual initiatives in single hospitals over past 10 years	Strong in a few leading hospitals based on individual leadership	Leading institutions, broad awareness through specialist societies

Source: Hofmann, A., Spahn, D. R., Holtorf, A. P., et al. (2021). Making patient blood management the new norm(al) as experienced by implementors in diverse countries. *BMC Health Services Research, 21*, 634. https://doi.org/10.1186/s12913-021-06484-3

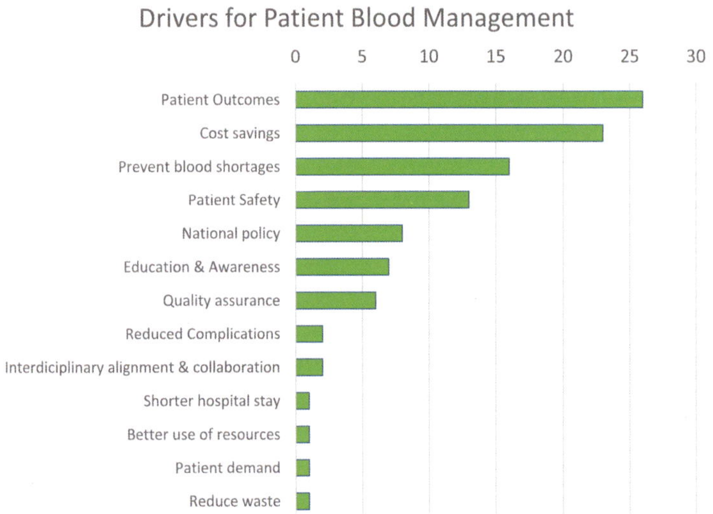

Figure 5.10. The drivers of patient blood management.

Source: Hofmann, A., Spahn, D. R., Holtorf, A. P., et al. (2021). Making patient blood management the new norm(al) as experienced by implementors in diverse countries. *BMC Health Services Research, 21*, 634. https://doi.org/10.1186/s12913-021-06484-3

Barriers	Australia	Brazil	China, PR	Croatia	Greece	Lebanon	Mexico	Saudi Arabia	Sth Africa	Sth Korea	Switzerland	Turkey	Average	STDev
Change of work practice	1.0	3.7	3.5	3.3	3.3	3.3	3.8	3.0	3.5	2.5	3.0	3.3	3.1	0.74
Collaboration / communication	4.0	3.0	4.0	2.5	3.5	3.3	1.8	3.3	3.0	2.8	2.0	3.0	3.0	0.69
No / limited PBM experience	3.0	3.0	2.0	2.5	2.7	3.0	2.5	3.3	3.3	3.3	4.0	2.5	2.9	0.52
Process / Feasibility	3.5	3.0	3.0	3.0	2.0	3.0	2.5	2.5	3.0	3.3	3.0	2.2	2.8	0.44
Strong belief in transfusion	0.5	2.7	1.0	3.5	1.8	3.0	4.0	2.8	3.0	3.8	3.0	3.0	2.7	1.06
Number of staholders	1.5	2.7	0.0	2.8	2.5	4.0	2.3	2.8	2.0	1.5	2.0	3.7	2.3	1.05
Cost	0.5	3.0	2.5	2.5	2.0	4.0	2.5	1.8	2.0	2.5	2.0	1.5	2.2	0.85
Sustainability	2.5	1.0	0.0	2.0	1.3	2.7	1.0	2.3	1.0	1.8	3.0	2.0	1.7	0.87
'Competition' w. other interventions	0.5	1.3	1.0	0.0	1.5	0.7	1.0	1.8	2.0	1.8	2.0	1.7	1.3	0.64

Figure 5.11. Barriers to implementing patient blood management.

Note:Between 0 (is no barrier) and 4 (high barrier). The barriers were sorted by the average rating (top to bottom decreasing). (N = 35). Color coding: red for average country values of 3 or higher, white for values between 2.01 and 2.99, and green for values of 2 or lower.

Source: Hofmann, A., Spahn, D. R., Holtorf, A. P., et al. (2021). Making patient blood management the new norm(al) as experienced by implementors in diverse countries. *BMC Health Services Research, 21*, 634. https://doi.org/10.1186/s12913-021-06484-3

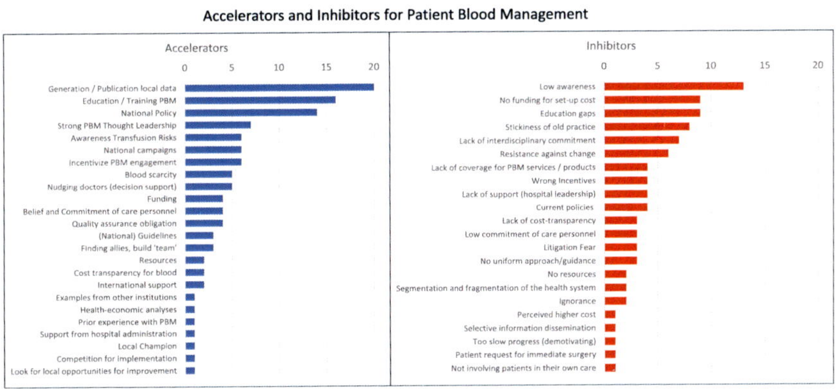

Figure 5.12. Accelerators and Inhibitors for patient blood management.

Source: Hofmann, A., Spahn, D. R., Holtorf, A. P., et al. (2021). Making patient blood management the new norm(al) as experienced by implementors in diverse countries. *BMC Health Services Research, 21*, 634. https://doi.org/10.1186/s12913-021-06484-3

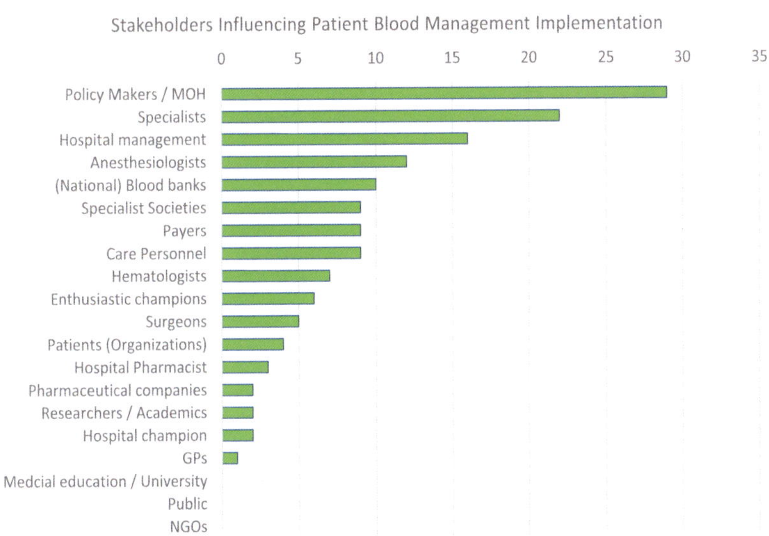

Figure 5.13. Stakeholders influencing patient blood management implementation.

Source: Hofmann, A., Spahn, D. R., Holtorf, A. P., et al. (2021). Making patient blood management the new norm(al) as experienced by implementors in diverse countries. *BMC Health Services Research, 21*, 634. https://doi.org/10.1186/s12913-021-06484-3

5.7. CASE STUDY 3: SUMMARY OF MEASURES TO CONSIDER FOR IMPLEMENTING PBM ACCORDING TO LEVEL OF INTERVENTION

Below are a summary of government (national and/or regional), funding, research, healthcare provision, training/education, and patients that can facilitate better blood management (Hofmann et al., 2021).

Table 5.19. A summary of measures suggested for better blood management by the PBM Implementation Group

Measures	Rational / Expectations	Examples	Points to Consider
Government Level (National or Regional)			
National Policy	• National initiatives and guidance push the hospitals towards PBM implementation	• AUS: National policy for PBM and national measures to support implementation • TUR, KOR: Close collaboration of PBM leaders with MoH • TUR: qualification for a 3-yr EU grant dedicated to a systematic country-wide implementation of PBM • CHN: officially addressed the importance of PBM to improve clinical practice	CHALLENGES • Changing of policy priorities; political instability (LBN) • Policy priority of shifting from tertiary hospitals to primary care level antagonizes the pre-operative PBM interventions (CHN) • National policy not a game changer in countries with decentralized healthcare (CHE)
Blood Shortage	• Actual and anticipated blood shortage is recognized on a policy level and requires action; donation volume is shrinking, the demand for blood is increasing (aging population) • Donor deferrals due to new or re-emerging pathogens; canceled blood collections due to lockdown during epidemics • Family replacement schemes: mandatory donations may increase risk and limit access	• ZAF, CHN, MEX: Frequent supply issues • HRV, GRE: Seasonal shortages • KOR, ZAF: Shortage and COVID-19 virus riska • BRA: Zika-Virus [57]; supply issues in public system • AUS, TUR: Shortage predicted • LBN, GRE, MEX: Replacement modus • CHN: 30% Reciprocal blood donation	REMARK • Impact of COVID-19 pandemic on blood supply
Strong PBM Thought Leadership	• Fosters a broader country-wide acceptance and change • Liaise on policy level, engage with payers, engage specialist societies, and introduce medical curricula	• KOR: Korean PBM Society with multi-disciplinary leadership function • TUR: EU-funded project for PBM implementation across Turkey	REMARK • May be difficult for few individuals to cover that scope and thus, formation of a high-level work or interest group may be advisable

PBM Incentives	• Attract clinicians to become part of the change • Increase level of experience and familiarity with PBM	• CRO: participation in international clinical study • TUR, KOR, MEX, ZAF: National pilots & research opportunities	REMARKS • Involve practitioners actively in research • Recognition of individual initiative through active engagement and authorship
National Guidelines	• Adapting international guidelines to local healthcare context can be essential for acceptance • National guidance will facilitate coordinated and homogeneous activities across the country	• ZAF, TUR, KOR, MEX, BRA, HRV, CHN: ongoing projects to locally adapt international guidelines	RISK • Scattered / fragmented approaches will make it difficult to consolidate in best practice
Funding Level			
Alignment of policy and funding	• Consensus for a reimbursement and funding solution	• KOR: Center for Disease Control in the MoH and the Health Insurance and Reimbursement Agency (HIRA) committed to PBM related projects (1) auditing the current level of transfusion appropriateness in each hospital, and (2) funding dedicated projects on PBM implementation in the country	CHALLENGE • Heterogeneity in access to healthcare and its funding requires different approaches for funding and reimbursement of PBM (MEX, LBN)
Reimbursement	• Increase the willingness to invest in establishing PBM • Adjust reimbursement systems to incentivize improved health outcomes and efficiency and disincentivize transfusion volume.	• KOR, MEX, AUS, TUR: Funding national pilot or full implementation projects • BRA: volume-dependent reimbursement to hospitals (fee for service) incentivizes a high use of transfusions; but first examples of capitation-based hospital reimbursement emerge (supportive for PBM)	CHALLENGES • Potential other sources of funding (NGOs, special international projects) • Funding always compromised during (economic) crises
Cost transparency for blood; Cost 'fairness'	• Mandate full cost transparency of transfusion and PBM to allow for cost-effective allocation of (public) funds	• GRE, HRV: Not knowing the cost of blood products or artificially low cost impedes adoption of PBM	REMARK • Even if at zero cost to the hospital, blood products are not for free from a societal perspective
Funding and resources in hospital	• Secure funding necessary for setting up the infrastructure (including point-of-care testing devices, cell salvage equipment, pre-operative anemia clinic, continuous medical education (CME) and training • Identify and remove dis-incentives	• HRV, GRE, LBN, MEX: Difficulties in securing funding despite principal support for the concept • BRA, LBN: Fee for service dis-incentivizes PBM (imbalance between profitability and patient health) • Alignment across budgets: e.g., pharmacy budget vs other cost	REMARK • Use measures / local data to demonstrate the realistic budget needs, ROI, time frame required .

Research Level			
Quality measurement/ assurance	• Use of quality measures, to track blood use (i.e., units ordered, used, and discarded per hospital, ward, type of intervention and individual specialists) to shift focus to patient needs and outcomes	• KOR, MEX, CHE, AUS: pursuing quality and performance measurement initiatives	REMARK • Performance measures empowers local transfusion committees and PBM implementation task forces
Collecting and publishing local data	• Demonstrating impact of PBM with local data on clinical outcomes, adverse events or complications, • Capturing and reporting local epidemiology data (prevalence) • Quantify opportunities, risks, and cost for PBM in the local setting; ideally as multi-disciplinary intra- or inter-hospital collaboration	• AUS, CHE, KOR: local data collection systems initiated or established to enable reporting, benchmarking, or performance analysis • TUR: publication of local data	REMARKS • Local evidence helps to refute that the international experience may not be transferrable to the local context • Local research motivates participants to gain expertise and to become part of the change
Health-economic analyzes	• To convince stakeholders of the cost-effectiveness of PBM, analyzes must be based on local data (cost / outcomes)	• Health-economic evidence from AUS, CHE, GER, and the USA	CHALLENGE • Current H/E evidence from countries with specific economic and health-economic settings and may not be generalizable
International support and collaboration	• Cross-fertilize and share the learnings transnationally	• International collaboration is frequent, e.g., strong engagement of IFPBM & SABM, ZAF w. National Blood Authority in AUS, KOR w. AUS, BRA w. SABM.	REMARK • Includes international teaching, web-based services, advisory exchange, or involvement of experts in another country's task forces.
Healthcare Provider Level			
Communication	• Strengthen belief and commitment of clinical staff • Re-align all stakeholders around the transfusion process	• GRE: Generation of an intra-hospital consensus and protocol with reporting system for restrictive blood use • MEX, ZAF, AUS: continuous communication, involvement, and feedback by coordinator / initiator in hospital, • ZAF, MEX: Chat-group in a social media platform to report local experiences, announce events, and post relevant publications, questions. and suggestions	A common vision and buy-in by those who need to change their practice is essential to achieve change

Identify allies, build teams	• To increase clout and trust across specialties • Foster multi-disciplinary collaboration, mutual endorsement and support	• LBN: Expanding across specialties already in initial phase added great impetus MEX, TUR: Multidisciplinary PBM Academies; LEB, KOR, ZAF: Multidisciplinary Iron Academies	REMARKS • PBM is a team effort • Supports forming a guiding coalition
Prior experience with PBM	• Expand the knowledge and openness for PBM by involving care personnel from different disciplines in implementation projects	• Pilot projects in several hospitals/wards to involve and expose them to PBM methods	REMARK • Overcome the stickiness of the old practice and resistance to change
Ensure support from hospital administration	• Design/align the organization to enable optimal and sustainable PBM across specialties • Secure funding for staff, systems support (IT), other resources • Get approval to establish a multi-disciplinary PBM committee	• Most initiatives reported that alignment with hospital administration / CEO was improving chances for success • HRV, GRE: Activities under departmental responsibility may not need agreement by hospital management. • LBN, SAU, HRV, LBN: To get funding for establishing PBM was difficult and therefore done within the existing resources (overtime) • BRA, MEX: dedicated project management ensures planning and roll out across specialties / departments	REMARKS • While small changes could be introduced individually or within one specialty the full potential can only be achieved with multi-disciplinary change • Understand the economic and system incentives and to be in close communication to collaboratively identify the path to implementation (milestones, tasks, and responsibilities)
Local champion (Medical Director or project coordinator for PBM)	• Responsible for planning, organizing and directing PBM, supporting specialists, and ensuring continuous data collection, reporting and benchmarking,	• HRV, GRE, MEX: general role in training, education, information, protocol development • BRA: Change management • AUS, CHE: organize PBM at patient level (case management)	REMARK • PBM coordinator can be a success factor for sustainability (AUS, CHE)
Hospital protocols (SOPs)	• Tailor PBM protocols to the specific hospital context and routines • Increase local ownership across the disciplines, interdisciplinary commitment	• HRV, GRE, TUR, MEX, BRA: Several interviewees reported the development of local protocols before the availability of National Guidelines	
Data collection, reporting & benchmarking system	• Shows impact, measures gaps, and helps to improve quality of care	• ZAF, KOR: currently developing a monitoring system in hospital(s)	

Nudging clinicians & stimulating competition	• Using IT or quality reporting systems to motivate and remind physicians to practice PBM • Using the competitive nature of people to motivate them to excel in PBM	• AUS, CHF, MEX, ZAF reported use or plan to use competitive forces or 'nudging instruments' to remind practitioners to improve their PBM practices (reminders, league tables)	REMARKS • Include IT and/or quality specialists in developing the local procedures for mapping into data collection and analytical support tools • udging=nonregulatory and nonmonetary interventions that steer people in a particular direction while preserving their freedom of choice"
Involving the entire care team	• Alignment, participatory processes	• ZAF: Importance of involving nurses who have high influence on the patient care • GRE: Importance of aligning the ordering of blood products.	REMARK • Includes nursing, hospital pharmacy, blood ordering process to ensure common goals
Seizing local opportunities for improvement	• Create momentum: Use opportunities in own environment for starting with specific aspects of PBM • Move forward faster and prove success	• HRV, LBN: Start within ward/ department • ZAF: start with communication & education of hospital specialists • MEX, BRA: pilots	REMARK • Even small 'wins' will motivate people
Training & Education Level			
Education and Training for PBM	• Identify and address knowledge gaps among specialists • Update under-and postgraduate curricula	• AUS: Integration in medical school (University of Western Australia) curriculum & exams • MEX, ZAF, TUR: PBM academies and/or continued medical education (CME) for practitioners • AUS, ZAF: online training material	REMARKS • Training of all specialists concerned (incl. anesthesiologist, intensive care specialists, surgeons, hematologists, oncologists, gastroenterologists, obstetricians & gynecologists) and nursing staff in relation to benefits of PBM, • Avoid asymmetry in information to prevent that 'eminence wins over evidence' in the choice of therapy
Increase Awareness Transfusion Risks	• Overcome eminence-based practice ("transfusion is always beneficial") and increase the knowledge about the associated risks	• Global: Many of the specialists who administer transfusions during surgery (surgeons, anesthesiologists) often don't see the mid- or long-term complications (infections, immune reactions, thrombosis).	REMARKS • Necessitates re-education of all participants in the transfusion decision

			• Requires information, education, and reminders across specialties (publications and newsletters, conferences, social media-channels)
Medico-legal aspects and protective measures as part of PBM training	• Strengthen the assertiveness of physicians relating to PBM	• BRA: Litigation is commonly used by patients to get access to procedures which they perceive to be beneficial	
Patient & Public Level			
National information campaigns	• Develop awareness for PBM • Encourage patients to discuss PBM at their doctor's appointment • Prevent litigation against physicians following guideline-compliant restrictive transfusion strategies • Decrease patient demand blood transfusion	• KOR, LBN, ZAF: Initiated or conducted national awareness campaigns through important media channels • BRA: Litigation is commonly used by patients to get access to procedures which they perceive to be beneficial	RISKS • If done too early, doctors might be overwhelmed by patient demand • Too much information on transfusion risks may negatively impact the willingness of the public to donate blood REMARK • Involving patients, collaborating with patients, and informing the public may improve understanding and reduce the risk for litigation
PAG initiatives	• Co-create national information campaigns (PBM thought leaders, politicians, PAGs) • Explore patient experiences and preferences • Engagement / advocating for PBM insurance coverage • Achieve comprehensive patient education on risks and benefits of all treatment options (including transfusion) for anemia, blood loss and coagulopathy • Ensure fully informed consent and/or shared decision making • PAGs to request PBM certification and/or hospitals accreditation	• HRV, KOR: Initial contacts • TUR: In contact with 5 NGO's, who receive regular information • GRE, LBN, HRV, KOR, BRA: increasing demand for participatory medicine and shared decision making by PAGs and/or healthcare policy	REMARKS • PAG-patient interaction relating to transfusion and/or PBM not yet common • Co-creation / co-production: researchers, practitioners and the public join efforts and share responsibilities to develop, implement, monitor, evaluate and re-develop interventions

Source: Hofmann, A., Spahn, D. R., Holtorf, A. P., et al. (2021). Making patient blood management the new norm(al) as experienced by implementors in diverse countries. *BMC Health Services Research, 21*, 634. https://doi.org/10.1186/s12913-021-06484-3

Abbreviations: MoH Ministry of Health, SABM Society for the Advancement of Blood Management (https://sabm.org), ROI Return on Investment, AUS Australia, BRA Brazil, CHN People's Republic of China, HRV Croatia, GRC Greece, KOR Republic of Korea, LBN Lebanon, MEX Mexico, ZAF South Africa, CHE Switzerland, TUR Turkey, PAG Patient Advocacy Group, IFPBM International Foundation Patient Blood Management, SABM Society for the Advancement of Blood Management.

The risk of COVID-19 viral infection only became apparent starting in January 2020. Hence, this threat was only mentioned in the last interviews (KOR, ZAF, SAU).

"A potent message for society and community leaders: blood donation is a life-saving act of solidarity that every healthy adult individual in line with national donor recruitment criteria can perform. Promoting donation of blood as a civic duty is a good approach for health leaders to promote community buy-in for extensive voluntary blood donation and the required grassroots engagement. It is unacceptable for any community to lack access to secure blood products and services, regardless of location.

World Blood Donor Day is commemorated to thank the altruistic acts of selflessness of voluntary blood donors and celebrate life and humanity. This also sets an example for potential new donors.

In this philanthropic act, every stakeholder at the sub-national, national, regional, and global levels must come together to invest in, strengthen, and sustain national blood programs and ensure the availability of safe blood and blood products to all who need it."

—Dr Poonam Khetrapal Singh, WHO Regional Director for South-East Asia

REFERENCES

1. Alghamdi, S. Y. (2023). A review of blood delivery for sustainable supply chain management (BSCM). *Sustainability, 15*(3), 2757. https://doi.org/10.3390/su15032757 [© 2023 by the author. Licensee MDPI, Basel, Switzerland. This article is an open access article distributed under the terms and conditions of the Creative Commons Attribution (CC BY) license: https://creativecommons.org/licenses/by/4.0/].
2. Bennett-Guerrero, E., Zhao, Y., O'Brien, S. M., et al. (2010). Variation in use of blood transfusion in coronary artery bypass graft surgery. *JAMA, 304*(14), 1568–1575. https://doi.org/10.1001/jama.2010.1406.
3. García-Erce, J. A., Romón-Alonso, Í., Jericó, C., Domingo-Morera, J. M., Arroyo-Rodríguez, J. L., Sola-Lapeña, C., Bueno-Cabrera, J. L., Juárez-Vela, R., Zalba-Marcos, S., Abad-Motos, A., et al. (2021). Blood donations and transfusions during the COVID-19 pandemic in Spain: Impact according to autonomous communities and hospitals. *International Journal of Environmental Research and Public Health, 18*(7), 3480. https://doi.org/10.3390/ijerph18073480.
4. Gill, S., Adenan, A. M., Ali, A., & Ismail, N. A. S. (2022). Living through the COVID-19 pandemic: Impact and lessons on dietary behavior and physical well-being. *International Journal of Environmental Research and Public Health, 19*(2), 642. https://doi.org/10.3390/ijerph19020642.
5. Hadimani, S., Das, S., R., K., mohiyuddin S.M, A., & K, P. (2022). Patient blood management and the Covid-19 Pandemic: An institutional study. *Asian Journal of Health Sciences, 8*(1), ID27. https://doi.org/10.15419/ajhs.v8i1.498.
6. Hofmann, A., Spahn, D. R., Holtorf, A. P., et al. (2021). Making patient blood management the new norm(al) as experienced by implementors in diverse countries. *BMC Health Services Research, 21*, 634. https://doi.org/10.1186/s12913-021-06484-3 [*Open Access: This article is licensed under a Creative Commons Attribution 4.0 International License, which permits use, sharing, adaptation, distribution and reproduction in any medium or format, as long as you give appropriate credit to the original author(s) and the source, provide a link to the Creative Commons license, and indicate if changes were made. The images or other third party material in this article are included in the article's Creative Commons license, unless indicated otherwise in a credit line to the material. If material is not included in the article's Creative Commons license and your intended use is not permitted by statutory regulation or exceeds the permitted use, you will need to obtain permission directly from the copyright holder.*

To view a copy of this license, visit http://creativecommons.org/licenses/by/4.0/. The Creative Commons Public Domain Dedication waiver (http://creativecommons.org/publicdomain/zero/1.0/) applies to the data made available in this article, unless otherwise stated in a credit line to the data.].

7. Juárez-Vela, R., Andrés-Esteban, E. M., Santolalla-Arnedo, I., Ruiz de Viñaspre-Hernández, R., Benito-Puncel, C., Serrano-Lázaro, A., Marcos-Neira, P., López-Fernández, A., Tejada-Garrido, C. I., Sánchez-González, J. L., et al. (2022). Epidemiology and associated factors in transfusion management in intensive care unit. *Journal of Clinical Medicine, 11*(12), 3532. https://doi.org/10.3390/jcm11123532 [© 2022 by the authors. Licensee MDPI, Basel, Switzerland. This article is an open access article distributed under the terms and conditions of the Creative Commons Attribution (CC BY) license: https://creativecommons.org/licenses/by/4.0/].
8. Rahmartani, L. D., Dewi, M. K., Iskandar, S. D., Pratanata, A. M., Ilmana, G., Sari, T. T., Lubis, A. M., &Wahidiyat, P. A. (2023). Impact of COVID-19 pandemic on pre-transfusion hemoglobin level and frequency of transfusion in transfusion-dependent thalassemia patients in Indonesia. *Thalassemia Reports, 13*(1), 1–9. https://doi.org/10.3390/thalassrep13010001 [© 2022 by the authors. Licensee MDPI, Basel, Switzerland. This article is an open access article distributed under the terms and conditions of the Creative Commons Attribution (CC BY) license: https://creativecommons.org/licenses/by/4.0/].
9. Rubin, R. (2022). COVID-19 and blood donation. *JAMA, 327*(7), 615. https://doi.org/10.1001/jama.2022.0763.
10. Salem-Schatz, S. R., Avorn, J., &Soumerai, S. B. (1990). Influence of clinical knowledge, organizational context, and practice style on transfusion decision making. Implications for practice change strategies. *JAMA, 264*(4), 476–483.
11. Tsante, A., Papandreadi, A., Tsantes, A. G., Kyriakou, E., Douramani, P., Loukopoulou, E., Gialeraki, A., Kokoris, S. I., Kypraiou, A., Poulis, A., et al. (2021). Seeking strategies to optimize blood utilization: The preliminary experience with implementing a patient blood management program in a Greek tertiary hospital. *Journal of Clinical Medicine, 10*(10), 2141. https://doi.org/10.3390/jcm10102141 [© 2021 by the authors. Licensee MDPI, Basel, Switzerland. This article is an open access article distributed under the terms and conditions of the Creative Commons Attribution (CC BY) license: https://creativecommons.org/licenses/by/4.0/].

12. Vasiljević-Jovanović, B., Milenković, M., Mijović, L., Bukumirić, Z., Šantrić-Milićević, M., Božanić, M., Milutinović, V., &Zdravković, M. (2021). Evaluation of the use of blood products in ICU hospitalized COVID-19 patients. *Surgeries, 2*(4), 391–398. https://doi.org/10.3390/surgeries2040039.

INDEX

I

L

M

R

S

T